INTRODUCTORY SOILS

LABORATORY MANUAL

J. J. Hassett

Department of Natural Resources and Environmental Sciences

University of Illinois

ISBN 0-87563-791-4

STIPES PUBLISHING L.L.C.
204 W. University Ave.
Champaign Illinois 61820

TABLE OF CONTENTS

LABORATORY 1.

SOIL COLOR AND TEXTURE-BY-FEEL.

1.1 SOIL COLOR.

Color is an easily determined soil characteristic. Even though soil color directly effects only the absorption of solar radiation, soil color can provide valuable information about other soil properties. The color of surface horizons is often related to their organic matter content. The color of subsurface horizons provides information about internal soil drainage. In addition, differences in soil color is one of the properties that can be used to differentiate soil horizons.

1.1.1 Munsell color notation system.

Soil color is most conveniently measured by comparison of the soil color with colors on a color chart. The collection of color charts generally used with soils is a modified version of the collection of charts appearing in the Munsell Book of Color and includes only that portion needed for soils. The Munsell notation identifies color by the use of three variables, Hue, Value and Chroma.

Hue is the dominant spectral (rainbow) color; that is whether the color is yellow, red, green or mixtures such as yellow-red. Mixtures are identified numerically according to the amount of yellow or red used to produce the mixture. 5YR is an equal mixture of red and yellow. As the number increases the amount of the first letter color (Y, yellow) increases and as the number decrease the amount of the second letter color (R, red) increases.

Value and Chroma are terms that refer to how the hue is modified by the addition of grey to the pure color (hue). Value is a property of the grey color that is being added to the hue. A particular grey (value) is made by mixing a pure white pigment (10) with a pure black pigment (0). If equal amounts of white and black pigments are mixed then the value is equal to 5, if more black than white pigment is used then the value is less than 5, if more white than black pigment is used then the value is greater than 5. Chroma is the amount of pure hue that is mixed with a grey of a particular value to obtain the actual color. A chroma of 1 would be made by adding one unit of pure hue to certain amount of grey, a chroma of 5 would contain 5 units of pure hue to that amount of grey. The lower the chroma the greyer the color.

1.1.2 Determination of soil color.

The nomenclature for soil color consists of two complementary systems: color names and the Munsell notation of color. Color names are less precise, but convey a general concept of the color of the soil. Munsell notation is more precise and is standardized so that soil scientists in different countries will have no difficulty in communicating information about soil color.

The Munsell soil color charts are set up so that each page is a separate hue, for example 10YR or 5R. The hue is given in the upper right hand corner of the chart (page). An actual chart is a collection of color chips of constant hue arranged by value and chroma. Rows of color chips are for a constant value, with chromas increasing from left to right. Columns are for a constant chroma with values decreasing from top to bottom.

Soil color is determined by matching a moist soil sample with the appropriate color chip. Because a given color chart is not a collection of all possible values and chromas of a given hue, the match is often less than perfect. It is generally possible to describe one chip that is very close in color to the moist soil. The major difficulties encountered in using the soil color charts are 1) in selecting the appropriate hue card 2) determining colors that are intermediate between hues and 3) distinguishing between values and chromas when chromas are strong (high numbers).

Once the appropriate color chip has been selected, the hue, value and chroma are recorded. The notation 5YR 5/4 is for a soil with a hue of 5YR, a value of 5 and a chroma of 4. The corresponding color name for 5YR 5/4 is reddish brown. Color names are often given on the opposite (facing) pages of the color books. If expression of soil color more precise than whole numbers for values and chromas is desired, decimals are used, never fractions. Generally color is only given in integer (whole) numbers. Reproducible measurements of soil color can be obtained at two moisture contents: moist (field-capacity) and air-dry. In most soil descriptions, unless otherwise stated, colors are given for moist soils.

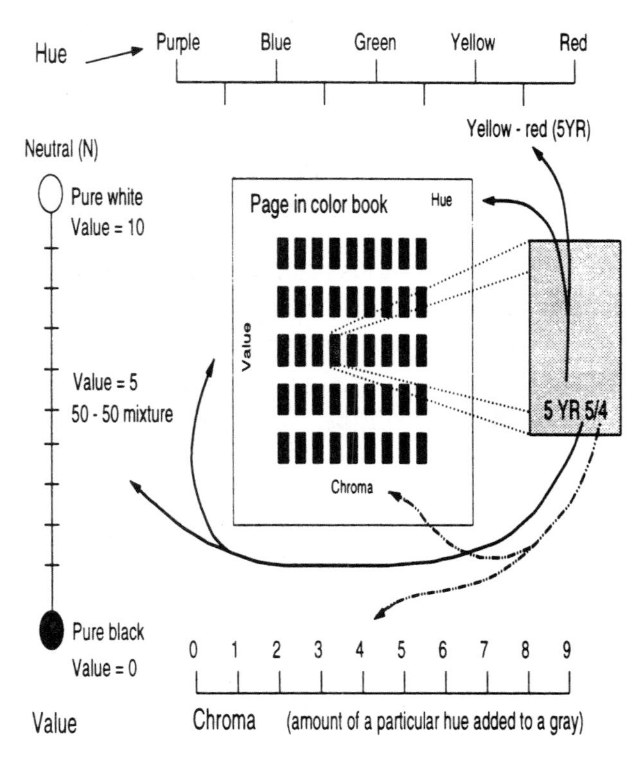

1.1.3 Interpretation of soil color.

Soil color is basically due to: 1) The presence of soil organic matter (humus). Organic matter imparts a brown to black color to the soil. Generally the higher the organic matter content of the soil the darker the soil. 2) The oxidation status of the iron compounds in the soil. In the lower horizons, where the soil minerals are not coated with humic substances, the color of the soil minerals will predominate. In better drained and, hence, well-aerated soils Fe(III) minerals give soils a red or yellow color. In more poorly drained and, hence, poorly-aerated soils the iron minerals are reduced and the neutral (grey) colors of Fe(II) minerals and the other soil minerals predominate.

Soil organic matter. The organic matter content of soils can be estimated from the Munsell color of a soil (J. D. Alexander, Univ. of Illinois AG-1941). The most accurate estimates are obtained with medium and fine textured mineral soils. Soils with greater than 50% sand and less than 10% clay usually contain less organic matter than predicted. The performance of many herbicides is influenced by adsorption onto soil organic matter. Estimation of soil organic matter from soil color can be helpful in selecting herbicides and determining application rates. Appropriate rates maximize weed control, while minimizing crop damage and potential environmental effects.

Table 1-1 Relationship of soil color to soil organic matter content

Munsell notation	percent organic matter	
moist color	range	avg
10YR 2/1	3.5 - 7.0	5.0
10YR 3/1	2.5 - 4.0	3.5
10YR 3/2	2.0 - 3.0	2.5
10YR 4/2	1.5 - 2.5	2.0
10YR 5/3	1.0 - 2.0	1.5

Soil drainage classes. The drainage class of a soil can be determined from the colors and color patterns in the soil's lower horizons (subsoil). The red color of soils is generally related to the presence of unhydrated iron(III) oxides, although manganese dioxide and partially hydrated iron(III) oxides may also contribute red colors. The red colors may be inherited from the parent materials or developed by the oxidation of iron minerals during soil weathering. Red colors are stable only in soils that are well-aerated.

The yellow color of soils is largely due to the presence of hydrated iron(III) oxides. Soils with yellow colors tend to occupy wetter landscape positions than associated red soils, this results in the hydration of the iron(III) oxides. Grey and whitish colors of soils are caused by several substances, mainly quartz, kaolinite and other clay minerals, calcium and magnesium carbonates (limestone) and reduced iron compounds. The greyest colors (chromas less than 1) occur in permanently saturated soil horizons, these soils often have a bluish appearance.

Soil horizons may be uniform in color or may be streaked, spotted, variegated or mottled. Local accumulations of carbonates or organic matter can produce a spotted appearance. Streaks or tongues of color may result from the downward movement of clays, organic matter and/or iron oxides. Mottling is often associated with fluctuating water tables creating changes in soil drainage and aeration and, hence, mixtures of red, yellow and grey colors.

Natural drainage classes.

Very poorly drained — Soils on level or depressional areas frequently ponded with water. Black or dark grey surface horizons, light grey colored horizons immediately under the surface horizons.

Poorly drained	Soils having high water tables or slowly permeable layers in the profile. Mottling occurs immediately under the surface horizons. Lower horizons light grey in color.
Somewhat poorly drained	Usually has mottles between 25-45 cm from soil surface. Light grey colors not presence, except deep into parent materials.
Moderately well drained	Usually mottling first occurs 45-75 cm from soil surface.
Well drained	Usually free from mottles, has uniform brown and yellowish or reddish brown colors in subsoil. If mottles occur they are below 75 cm depth.

1.3 SOIL TEXTURE.

Soil texture refers to the percentage of sand, silt and clay particles in a given mass of soil. Particles greater than 2 mm in diameter are removed from the soil by sieving and are excluded from the textural determination. The presence of larger particles is recognized by the use of modifiers such as gravelly, cobbly, stoney, cherty, slaty or shaly based on the size and composition of the larger particles.

Sand Particles ranging in size from 2.00 mm to 0.05 mm in diameter.

Silt Particles ranging in size form 0.05 mm to 0.002 mm in diameter.

Clay Particles less than 0.002 mm (2 μm) in diameter.

Sand, silt and clay are size separates that include all mineral particles in a specific size range regardless of composition or mineralogy.

1.3.1 Textural classes.

Soils normally are a mixture of sand, silt, clay, larger mineral particles and organic matter. The larger mineral particles and the organic matter are removed before textural analysis. Soils are then classified into different textural classes based on the percentage of sand, silt and clay in the soil. Textural classes share similar physical and chemical properties, that is similar water holding capacities or cation exchange capacities, etc.

Determination of a soils textural class is best achieved using a textural triangle. A textural triangle is constructed so that the bases represent 0% sand or silt or clay and the corresponding apexes represent 100% sand or silt or clay.

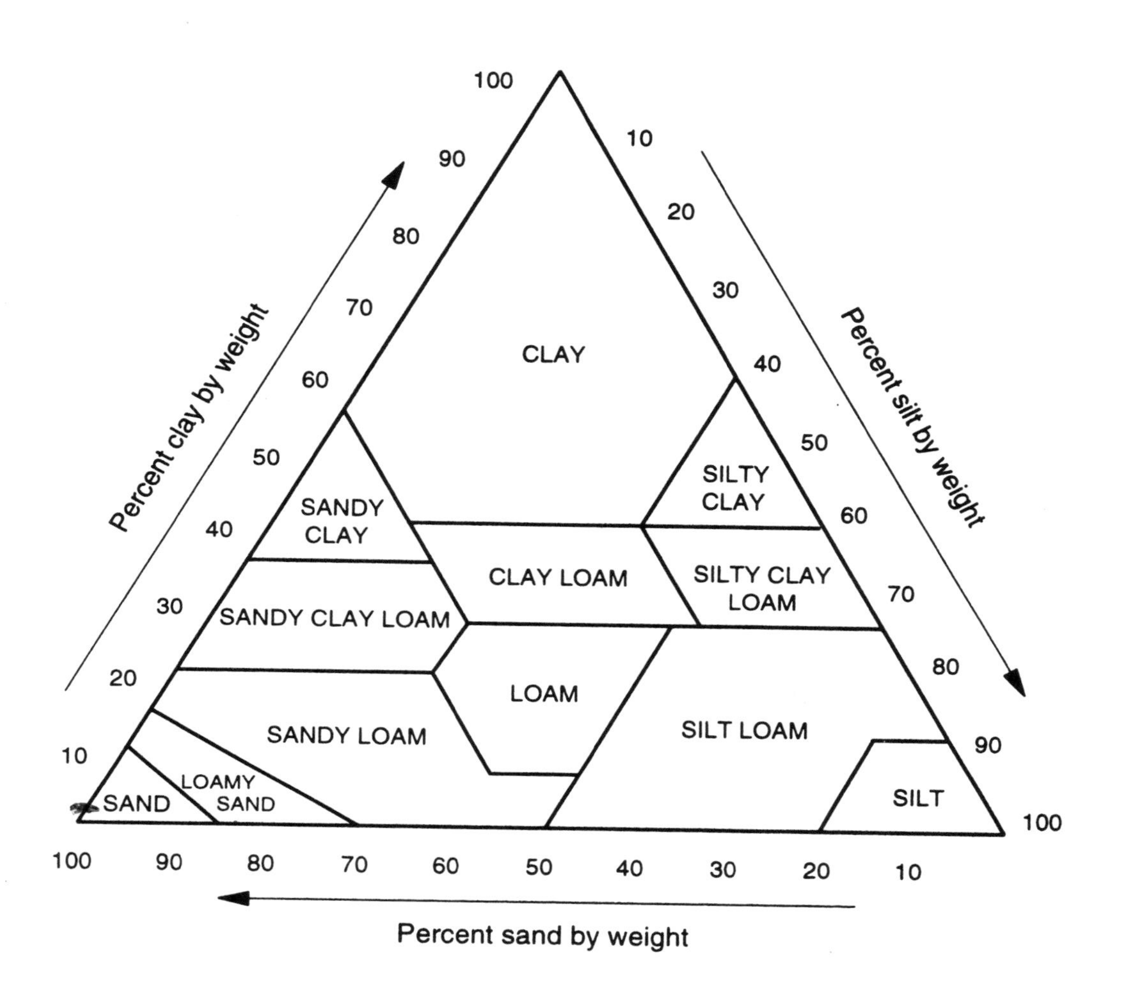

100
90
80
70
60
50
40
30
20
10
Percent clay by weight
10
20
30
40
50
60
70
80
90
100
Percent silt by weight
CLAY
SILTY CLAY
SANDY CLAY
CLAY LOAM
SILTY CLAY LOAM
SANDY CLAY LOAM
LOAM
SANDY LOAM
SILT LOAM
SILT
LOAMY SAND
SAND
100
90
80
70
60
50
40
30
20
10
Percent sand by weight

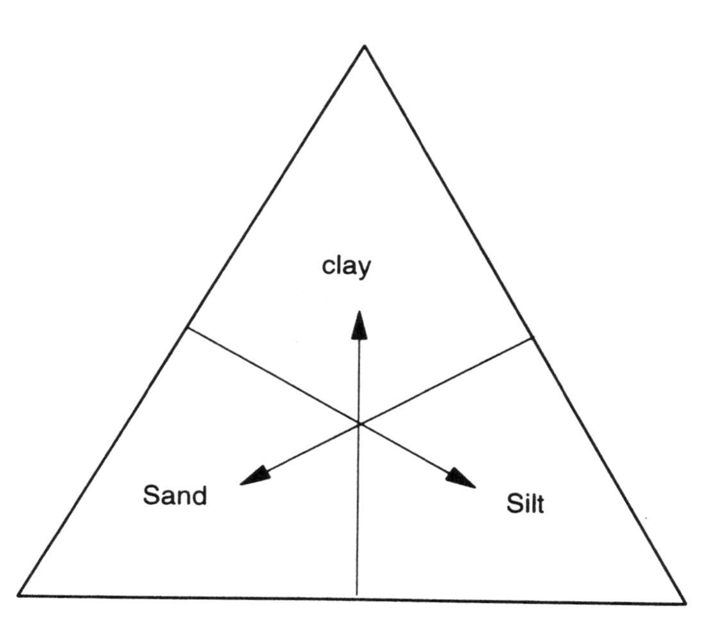

clay
Sand
Silt

Laboratory directions.

1. Determine the soil color of the unknown samples provided by your instructor. Practice and develop your technique with the known samples. Estimate the organic matter content of the unknowns.

   Procedure for determining soil color.

   a. Select the ped or soil sample to be described.

   b. Moisten the sample so that no free water is present nor is it glistening in the light.

   c. Determine the hue of the sample by comparison with pages in the Munsell color book. Begin with the 10YR hue as many midwestern soils have this hue.

   d. With good light coming over your shoulder compare the soil's color with the color chips until you find the best match possible. You may have to select another hue (page) if you have difficulty finding a match.

   e. Record the Munsell color notation (hue value/chroma) and the soil color name.

2. Determine the texture of the unknown samples provided by your instructor using the grittiness and ribbon tests. Develop your technique using the known samples.

   Procedure for determining texture-by-feel.

   a. The grittiness test measures the presence of sand or silt in the soil sample. The soil sample is placed in the palm of the hand, excessively wet and then rubbed with the forefinger of the opposite hand. Sand feels gritty, silt feels silky (smooth).

b. The ribbon test measures clay content by determining the strength and length of a ribbon of soil that can be formed when a properly wet soil sample is squeezed between the thumb and forefinger of the same hand. The higher the clay content the stronger the ribbon.

3. Data - soil color, organic matter estimate and texture by feel

	Soil 1	Soil 2
Color (Munsell notation)	_______	_______
Estimated organic matter content	_______	_______
Grittiness test (low, med or high sand)	_______	_______
Ribbon test (low, med or high clay)	_______	_______
Textural class	_______	_______

4. Next week hand in:

   a. The data for the color, organic matter content, and texture of the unknown samples on a separate sheet (i.e., leave the data sheet in your lab notebook).

   b. A one page discussion on the significance of soil color and texture. You should emphasize the type of information that these soil properties provide about a soil.

LABORATORY 2.

SOIL TEXTURE (HYDROMETER METHOD), BULK AND PARTICLE DENSITY.

Texture, bulk and particle density are physical properties of soils that control many important soil processes. Texture affects the total water holding capacity of the soil, percentage of plant-available water, cation exchange capacity and many other soil properties and processes. Bulk and particle density are related to soil porosity, degree of compaction, movement of air and water into and through the soil, ease of root growth as well as other properties.

2.1 HYDROMETER METHOD FOR SOIL TEXTURE.

The determination of the size distribution of soil particles is known as mechanical or particle size analysis. Soil texture is the composition of the soil particles expressed as the percent of particles in the sand, silt and clay size separates after organic matter, carbonates and iron and manganese oxides and other cementing or binding agents are removed.

The hydrometer method[1] is based on the change of density of a soil and water suspension upon the settling of the soil particles. Stokes' Law is used to predict the settling times for various sized particles. Stokes' law states that the rate which particles fall in a viscous medium (water) is governed by the radius of the particles and the force due to gravity. A special hydrometer, calibrated in terms of the grams of soil **suspended**, is used to measure density. The hydrometer is gently placed into the cylinder containing the suspension after predetermined periods of time and a reading taken by determining where the meniscus of the suspension strikes the hydrometer.

2.1.1 Stokes Law.

The rate of fall (v) of a particle in a suspension can be predicted from Stokes' Law:

$$v = 2r^2(\rho_s - \rho_l)g/9n$$

Where:
- v = velocity of particle falling in a liquid, cm/sec
- r = radius of particles, cm
- ρ_s = density of the solid particles, ~2.65 g/cm^3
- ρ_l = density of the liquid, g/cm^3
- g = acceleration due to gravity, 980 cm/sec^2
- n = viscosity of the liquid, poises

---

[1]. Bouyoucos G.J. 1962. Hydrometer method improved for making particle size analysis of soil. Agron. J., 54:464-465.

The determination of texture by the hydrometer method is based on certain assumptions:

1. That soil particles are spherical and are large enough so that Brownian movement does not affect their settling.

2. That soil particles are of identical density.

3. That the particles fall independently, there is no interaction between particles.

4. There are no temperature gradients or currents to affect the density and viscosity of liquid.

The first assumption that soil particles are spherical is not always valid since many soil particles are plate-like. Because of this Stokes' Law is used to calculate an approximate settling time and then the time is adjusted to match settling times for known soil size separates. The assumption of identical density is reasonable since most soil minerals are silicate minerals and have similar densities. To insure that particles fall independently, cementing agents, such as organic matter, carbonates and iron oxides are removed and then a chemical dispersant (sodium hexametaphosphate) in combination with mechanical dispersion is used to separate the soil aggregates into individual particles. And last, temperature control can be obtained by placing the sedimentation cylinders in a temperature controlled water bath.

Laboratory directions - hydrometer method.

Some soil samples require that organic matter be removed by oxidation with hydrogen peroxide (H_2O_2), that carbonates be dissolved using a pH 5 acetate buffer and that iron compounds be removed by reduction with sodium dithionite. **For the purpose of this laboratory these procedures are omitted**.

a. Weigh approximately 50 grams of air dry soil (100 grams for sandy soils) that has passed a 2 mm sieve and quantitatively transfer into metal dispersing cup. Calculate the oven-dry weight of your soil sample using the air-dry water content (%) provided by your laboratory instructor.

b. Add 20 mL of 2.5 N sodium hexametaphosphate, fill to within two inches of the top of the dispersing cup with deionized water and let stand for 10 minutes.

c. Carefully attach the dispersing cup to the mixer and stir for 5 minutes.

d. Quantitatively transfer the dispersed sample (soil and solution) from the dispersing cup into a sedimentation cylinder.

e. Fill the cylinder with deionized water to the 1000 mL mark.

f. Calibrate the hydrometer used by placing it in a sedimentation cylinder that contains 20 mL 2.5 N sodium hexametaphosphate and 980 mL deionized water. This calibration procedure is necessary since not all of the hydrometers read 0 g/L when there is no soil in suspension. Subtract this value from the hydrometer readings if the calibration reading is greater than 0 and add this value to the hydrometer readings if the calibration reading is less than 0. **Be certain to use the same hydrometer through out the experiment or calibrate each new hydrometer that you use.**

g. Place a rubber stopper in the end of the cylinder and agitate vigorously by turning end to end. When all the soil material is resuspended set the cylinder down and record the exact time.

h. Immediately, **very carefully,** insert the hydrometer into the suspension. 40 seconds after the cylinder was set down record the hydrometer reading. Repeat steps (g) and (h) three times, use the average of these values in your calculations. The 40 second reading gives the amount of silt and clay still suspended after the sand particles have settled.

i. Measure and record the temperature of the suspension after both hydrometer readings (40 seconds and 2 hours).

j. At the end of the 2 hour settling period carefully place the hydrometer into the suspension and record the reading. This reading gives the grams of clay per liter still in suspension.

k. Pour the suspension into the crocks in the sinks (to avoid clogging the sinks) and clean the sedimentation cylinder and your work area.

Data sheet for hydrometer experiment.

	Soil 1	Soil2
Air-dry weight	103	____
Air-dry moisture content	____	____
Hydrometer calibration ____		11.2
Hydrometer reading (40 seconds)	10	____ 93
Temperature (40 seconds)	23°C	____
Hydrometer reading (2 hours)	-1 -.2	____
Temperature (2 hours)	22°C	____

Calculations for hydrometer method.

a. Determine the oven-dry weight of the soil sample. The laboratory instructor will provide the percent moisture of the air-dry soil samples.

Oven-dry wt. = Air-dry wt./(1 + decimal of % H_2O)

For example if the soil contains 2.2% water when air-dry.

Oven-dry wt. = Air-dry wt./(1.022)

b. Calculate the temperature correction for the hydrometer readings.

1. For each degree above 20° C add 0.4 g/L to hydrometer reading.

2. For each degree below 20° C subtract 0.4 g/L from hydrometer reading.

c. Using the calibrated and temperature corrected hydrometer readings and the oven-dry weight of the soil sample calculate the percent sand, silt and clay.

1. Grams of sand = oven-dry wt. - corrected 40 sec. reading.

2. Grams of silt + clay = corrected 40 sec. reading.

3. Grams of clay = corrected 2 hr. reading.

4. Grams of silt = corrected 40 sec. reading - corrected 2 hr. reading.

% sand = (grams sand/oven-dry wt.) x 100

% silt = (grams silt/oven-dry wt.) x 100

% clay = (grams clay/oven-dry wt. x 100

d. Use the texture triangle to determine the texture of your soil sample.

2.2 THE BULK AND PARTICLE DENSITIES OF SOIL.

The **particle density** of a soil is the average density of the solids. Since soils are primarily composed of silicate minerals this is a fairly constant value from soil to soil. Particle density will only vary when the is a marked change in soil mineralogy. Soils tend to have particle densities (ρ_s) close to 2.65 g/cm^3.

ρ_s = oven-dry wt./volume of the soil solids = OD wt./V_s

Where V_s is the volume of the soil solids.

Bulk density is the density of the soil (solids and pores). It differs from particle density in that the volume of the pores is included in the calculation. Bulk density varies from soil to soil and from horizon to horizon and is primarily a function of the amount of pore space in the soil.

ρ_b = oven-dry wt./total volume of the soil = OD wt./V_t

Where V_t is the total volume of the soils and is the sum of the volume of the pore space (V_p) and solid space ($V_t = V_s + V_p$).

The percent pore space of a soil can be related to bulk and particle densities by the following relations.

$$\%PS = V_p/V_t \times 100$$

$$\%PS = (V_t - V_s)/V_t \times 100 \qquad \text{since } V_p = V_t - V_s$$

$$\%PS = (1 - V_s/V_t) \times 100$$

since V_s = OD wt./ρ_s and V_t = OD wt./ρ_b

$$\%PS = (1 - \rho_b/\rho_s) \times 100$$

The percent pore space in a soil is a function only of bulk density (ρ_b) since particle density is a constant for a given horizon or soil. When ρ_b increases, percent pore space decreases. There is also a change in the distribution of micro to macro pores with changes in bulk density. Increases in bulk density, which are usually caused by compaction, tend to destroy the large macro pores in preference to the smaller micro pores. Since the macro pores are those pores that drain free of water and hence serve as the pathways for air and water movement, compaction tends to increase problems associated with excess water (perched water tables) and associated with poor aeration.

Laboratory directions bulk and particle densities.

1. Coated-clod method.

Determine the bulk density of a soil aggregate using the coated aggregate technique. A mass of soil (aggregate or ped) is removed from the soil profile without changing its natural structure. The aggregate is weighed, coated with a sealer and the weighed again. The aggregate is then suspended in water and weighed. Archimedes Principle is used to determine the volume of the aggregate. Once the original moisture content of the aggregate is determined the bulk density (ρ_b) of the aggregate can be determined. Archimedes Principle states that an object placed in a liquid is buoyed up by a force equal to the weight of the displaced liquid. Since the displaced liquid equals the volume of the object and water has a density of 1 gram per cubic centimeter, the increase in weight when the aggregate is suspended in water is equal to the aggregate's volume.

a. Select a natural aggregate, approximately 4-10 cm in diameter, from those provided by your laboratory instructor. Preserve natural structure as much as possible.

b. Weigh the air-dry aggregate. Record the moisture content of the aggregate (provided by instructor).

c. Tie a thin thread around the aggregate, leave a loose end ~50 cm long. Attach a piece of masking tape to thread for identification.

d. Weigh aggregate and string and record weight.

e. Dip aggregate into saran mixture (1:6 mixture by weight of saran to solvent). Remove and allow to drip so that excess falls into the saran container. When dripping slows, dip the aggregate a second time.

f. Allow to dry by hanging from thread on a ring stand for at least 5 minutes.

g. Examine the coated aggregate. If areas exist that are not sealed, dip the aggregate again until it is completely sealed and provides a water tight coating.

h. After the coating has dried, weigh the coated aggregate and record weight.

i. Place a plastic beaker on the top loading balance. Add enough water so that the coated aggregate can be completely immersed without touching the bottom or sides of the beaker. There must be an enough free space above the water level so the when the coated aggregate is immersed water will not flow out of the beaker. Record the weight of the beaker and the water.

j. Suspend the coated aggregate in the water so that it is completely immersed without touching the sides or bottom of the beaker. Tie the string to the ring stand so that it is steady and record the weight.

k. Clean and dry balance area when you are through.

<u>Data sheet for coated-clod method</u>.

Air-dry weight of aggregate _____

Weight of air-dry aggregate and thread ____

Weight of coated aggregate and thread ____

Weight of beaker and water (0 if tarred) ____

Weight of beaker, water and suspended aggregate ____

<u>Calculations for the coated-clod method</u>.

a. Oven-dry wt. of the aggregate = air-dry wt. / (1 + decimal % of water content).

b. Wt. saran coating = wt. coated aggregate - wt. air-dry aggregate and string.

c. Volume of saran coating = wt. saran coating / density of saran (density of saran = 1.3 g/cm^3).

d. Volume of aggregate + saran coating = (wt. coated aggregate + beaker + water) - (wt. beaker + water).

e. Volume of aggregate = (volume aggregate + saran coating) - (volume of saran coating).

f. Bulk density of the aggregate = OD wt. of the aggregate / volume of aggregate.

2. Approximate bulk and particle densities.

Determine the bulk and particles densities and the percent pore space in the two soil samples provided. The approximate methods are based on artificially packing soil into a known volume, determining the weight of the volume of soil and using the data to calculate the soil's bulk density. The soil is then placed in a graduated cylinder approximately half full of water and stirred to expel the air in the soil. The change in volume caused by the addition of the soil is equal to the volume of the soil solids. The volume of the soil solids along with the weight of the soil can then be used to calculate particle density. Once ρ_b and ρ_s are known the percent pore space in the soil sample can be calculated.

a. Determine the weight of a 25 mL graduated cylinder. (Note: 1 mL = 1 cm^3)

b. Fill to the 25 mL mark, by adding ~ 5 mL additions of soil and tapping lightly to pack the soil. Note: Vt for soil = 25 cm^3

c. Determine the weight of the graduated cylinder + soil and by difference the weight of the soil. Correct the soil's weight for moisture content.

d. Fill a 100 mL graduated cylinder with tap water to the 50 mL mark. Quantitatively transfer the soil from the 25 mL graduated cylinder to the water and stir to expel the air. Let stand ~ 5 minutes.1

e. Determine the change in volume resulting from the addition of the soil.
Note Vs = volume change (volume after adding soil and stirring - 50 mL)

Data sheet for approximate bulk and particle densities.

	Soil 1	Soil 2
Total volume of soil (Vt)	25 mL	25 mL
Weight of empty graduated cylinder	_______	_______
Weight of cylinder and 25 mL soil	_______	_______
Volume of water in 100 mL cylinder	_______	_______
Volume of soil and water	_______	_______

Calculations for approximate bulk and particle densities

Oven-dry weight	OD wt = air-dry weight/(1 + water content)
Bulk density	ρ_b = OD wt/Vt Note Vt = 25 cm^3
Volume of solids (Vs)	Vs = (volume of soil + water) - (volume of water)
Particle density	ρ_s = OD wt/Vs
Percentage pore space	%PS = $(1 - \rho_b/\rho_s) \times 100$

Next week hand in:

a. The data and calculations for the hydrometer, coated aggregate and approximate experiments.

b. A one page discussion of the experiments and the significance of the physical properties that they are designed to measure.

LABORATORY 3.

DESCRIPTION OF THE SOIL PROFILE.

A variety of information can be determined by a visual inspection of fresh and fixed cores in the laboratory. The profile is describe by identifying the individual horizons that are present, their depth, color, structure, and texture. Other information about the soil which may also be determined from a visual inspection includes; parent material, natural drainage class and taxonomy.

3.1 SOILS HORIZONS.

1. **Depth** (cm) - The mineral surface is assigned a depth of zero. Record depth to the center of the boundary separating each horizon from the one below, i.e., 0 - 15 cm, 15 - 30 cm, etc. Horizons thinner than 7.5 cm are not ordinarily described with the notable exceptions of E, Bh and Bs horizons.

2. **Texture** - Standard name and abbreviations are used to describe the texture of each horizon.

sand - - - - s	silt - - - - si	silty clay - - - - sic
loamy sand - ls	silt loam - sil	sandy clay loam - scl
sandy loam - sl	clay loam - cl	sandy clay - - - - sc
loam - - - - l	clay - - - - c	silty clay loam - sicl

3. **Color** - Use the Munsell soil color book to determine the moist color of each horizon. Record the color using the hue value/chroma notation.

   a. **Mottling** Mottling is the presence of more than one color due to impeded drainage.

4. **Horizon name** - Refer to the list and definitions of the master horizons in laboratory 1.

3.2 PARENT MATERIAL.

Parent Material refers to the material(s) from which the soil profile has been developed. More than one type of parent material is possible. This is shown with an arabic number in front of the horizon symbol, i.e., 2Bt etc.

1. **Fluvium (Recent alluvium)** - Fluvial material that has been transported by streams and deposited on present day flood plains or stream terraces.

2. **Outwash (Old alluvium)** - Fluvial material deposited by glacial meltwaters.

3. **Glacial till** - Nonstratified, nonsorted glacial material deposited by glaciers in ground, lateral and terminal moraines.

4. **Lacustrine sediments** - Relatively fine, well sorted, stratified materials deposited in fresh water lakes.

5. **Residuum** - Unconsolidated bedrock weathered in place, no transporting agent involved.

6. **Colluvium** - Material deposited on footslopes primarily by the action of gravity.

7. **Eolian sand** - Sand accumulated through the action of wind into a dune type topography.

8. **Loess** - Wind deposited silt sized material.

9. **Marine** - Lacustrine like material deposited in oceans and seas.

3.3 SOIL DRAINAGE CLASSES.

The natural drainage class and hence, aeration status of a soil can be fairly accurately determined from the colors and color patterns in the subsoils. The red color of soils is generally related to the presence of unhydrated iron(III) oxides (Fe_2O_3 hematite), although manganese dioxide and partially hydrated iron(III) oxides may also contribute red colors. The red colors may be inherited from the parent materials or developed by the oxidation of iron minerals by soil weathering processes. Red colors are stable only in soils that are well aerated.

The yellow color of soils is largely due to the presence of hydrated iron(III) oxides ($Fe_2O_3 \bullet 3H_2O$ limonite). Soils with yellow colors tend to occupy moister landscape positions than associated red soils, this result in the hydration of the iron(III) oxides.

Grey and whitish colors of soils are caused by several substances, mainly quartz, kaolin and other clay minerals, calcium and magnesium carbonates (limestone) and reduced iron compounds. The greyest colors (chromas less than 1) occur in permanently saturated soil horizons, these soils often have a bluish appearance.

Soil horizons may be uniform in color or may be streaked, spotted, variegated or mottled. Local accumulations of carbonates or organic matter can produce a spotted appearance. Streaks or tongues of color may result from the downward movement of clays, organic matter and/or iron oxides. Mottling and variegation is often associated with fluctuating water tables creating changes is soil drainage and aeration.

Natural drainage classes.

Very poorly drained	Soils on level or depressional areas frequently ponded with water. Black or dark grey surface horizons, light grey colored horizons immediately under the surface horizons.
Poorly drained	Soils having high water tables or slowly permeable layers in the profile. Mottling occurs immediately

	under the surface horizons. Lower horizons light grey in color.
<u>Somewhat poorly drained</u>	Usually has mottles between 25-45 cm from soil surface. Light grey colors not presence, except deep into parent materials.
<u>Moderately well drained</u>	Usually mottling first occurs 45-75 cm from soil surface.
<u>Well drained</u>	Usually free from mottles, has uniform brown and yellowish or reddish brown colors in subsoil. If mottles occur they are below 75 cm depth.

Laboratory directions.

1. Complete a profile description sheet for each of the fresh soil cores in the laboratory.

2. Answer the following questions:

   a. How would recent fluvial materials differ from glacial outwash, lacustrine sediments, glacial till and loess.

   b. What would be the probable texture of a soil forming at the center of a dry lake bed?

   c. Identify the most common parent materials in:

      Northeastern Illinois.
      Center Illinois.
      Southern Illinois close to horseshoe lake (Mississippi River flood plain).
      South central Illinois in the vicinity of Vandalia.

   d. What would be the drainage class of a soil if the B horizon contains mottling and is gleyed?

   e. How would you identify a well drained subsoil.

Introductory Soils

Profile Description Sheet

Profile no._____

Horizon	depth	texture	structure	color	mottles (Y/N)
____	______	____	_____	___	____
____	______	____	_____	___	____
____	______	____	_____	___	____
____	______	____	_____	___	____
____	______	____	_____	___	____
____	______	____	_____	___	____
____	______	____	_____	___	____

Parent Material ________________.

Natural Drainage Class ____________________.

Profile no.____

Horizon	depth	texture	structure	color	mottles (Y/N)
____	______	____	_____	___	___
____	______	____	_____	___	___
____	______	____	_____	___	___
____	______	____	_____	___	___
____	______	____	_____	___	___
____	______	____	_____	___	___
____	______	____	_____	___	___

Parent Material _______________.

Natural Drainage Class ___________________.

Profile no.____

Horizon	depth	texture	structure	color	mottles (Y/N)
____	______	____	_____	___	___
____	______	____	_____	___	___
____	______	____	_____	___	___
____	______	____	_____	___	___
____	______	____	_____	___	___
____	______	____	_____	___	___
____	______	____	_____	___	___

Parent Material ________________.

Natural Drainage Class ____________________.

LABORATORY 4

SOIL FIELD TRIP.

Examination of soils in the field provides information about the soil in addition to the profile description. Slope and landscape position often determines the soil's internal drainage, the thickness of the surface horizons relative to other soils in the landscape, as well as the sequence of horizons found in the soil profile. Soils that occupy the same type of landscape position and slope often share similar properties and are classified the same even when separated by great distances. In Illinois, somewhat-poorly drained soils are often found on flat level areas that are basically in equilibrium with precipitation, that is these site are not receiving nor losing large amounts of water due to runon of runoff of precipitation from other site positions. Soils forming on sloping sites that lose water due to runoff are generally well drained and moderately-well drained, while soils occurring in depression areas or other sites that receive water that is running off other landscape positions are generally poorly or very-poorly drained.

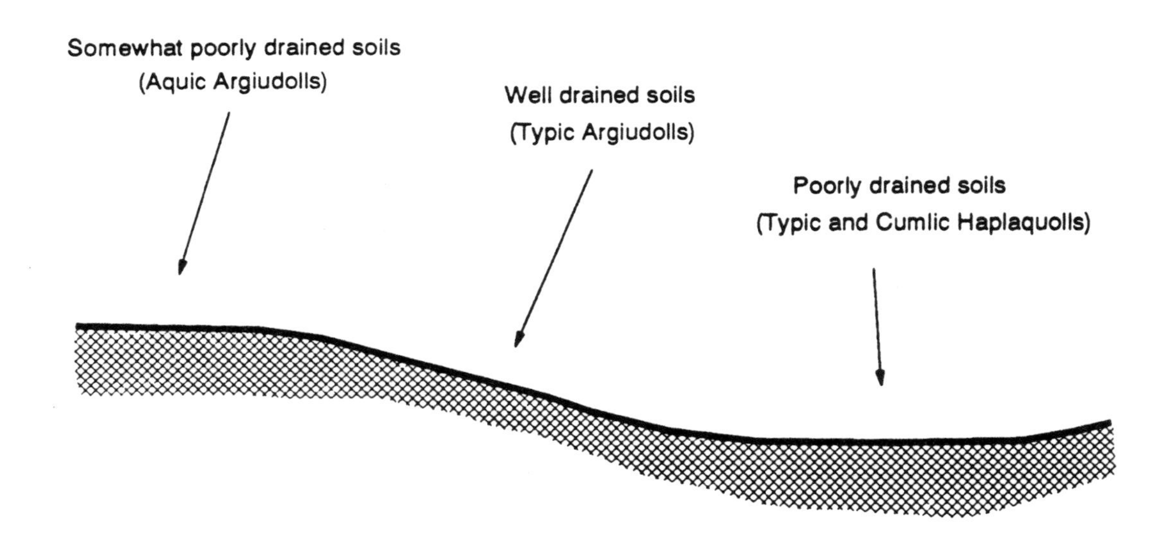

4.1 SITE POSITION

This category describes the landform on which the site is located. Site position is often closely related to the parent material from which the soil profile developed, i.e., fluvial material in a flood plain.

1. **Flood plain** - The lowest level of a stream valley, often referred to as the first bottom.
2. **Stream terrace** - Old bottoms of present or past streams higher in the landscape than the flood plain.

3. **Upland** - Other landforms on which soils are usually developed in loess, glacial till, glacial outwash, or bedrock.

4. **Foot slope** - This refers to the position at the base of a slope where colluvial material accumulates.

5. **Lake plain** - Sediments deposited in the bottom of past lakes. Material tends to be very fine and sorted with the coarse material near the old shore line and the finer material near the center.

4.2 SLOPE.

Slope is determined using an Abney level. It is usually determined by sighting from one stake to a second stake of the same height a few meters away. Slope is related to the degree of erosion of a soil, the amount of runoff and potential for ponding. The better drained soils in the landscape generally occur on sloping sites, while the worst drained soils in the landscape most often occur at the base of slopes or in depression areas.

4.3 SOIL TAXONOMY.

The modern soil taxonomy system (USDA) is based on the classification of soils from the properties that occur in the field. Pedogenic processes effect the classification of soils in the new system since they determine many of the soil properties, but they are not the dominate differentiating property that they were in many of the older systems.

Soils are classified by the presence or absence of special types of pedogenic horizons called diagnostic horizons. For example, soils are classified into different orders based on the type of A or B horizons present.

Diagnostic surface horizons.

Diagnostic surface horizons are called epipedons, meaning the surface of the soil. These are special types of A horizons.

1. **Mollic epipedon** - A thick dark colored high base saturated (> 50%) mineral surface (A) horizon. Specifically this horizon has sufficient soil structure that it is not massive nor hard when dry. The colors include chromas and values of 3.5 or darker when moist and values darker than 5.5 when dry.

The mollic epipedon must be more than 10 cm if lying directly over hard rock. If the soil contains a B horizon (i.e., an argillic, natric, cambic or spodic diagnostic horizon) or a fragipan or durapan, the mollic epipedon must be at least 1/3 the thickness of the solum when the solum is less than 75 cm thick and at least 25 cm thick when the solum is thicker than 75 cm.

The mollic epipedon must contain less than 30% organic matter or it becomes a Histic epipedon. It must also contain less than 250 ppm of citrate extractable P_2O_5 or it becomes an Anthropic epipedon.

2. **Histic epipedon** - Strongly developed organic (O) horizons of mineral soils. The horizon is from 20 to 60 cm thick. Usually formed on soils that under natural conditions are saturated with water a least 30 days per year. Organic carbon content ranges from 12 to 18% as a minimum depending upon clay content.

3. **Umbric epipedon** - Epipedons that meet all the requirements of a mollic epipedon except that they are too low in base saturation (i.e., too acidic).

4. **Ochric epipedon** - Surface (A or O) horizons that are too light in color, too low in organic matter or too thin to be a mollic, umbric, anthropic, plaggen or histic epipedon.

5. **Anthropic epipedon** - Surface horizons with the characteristics of a mollic epipedon, but formed due to the activities of man. Extractable P_2O_5 is > 250 ppm due to anthropogenic activities.

6. **Plaggen epipedon** - Man made surface horizon, produced by long term manuring, often thickened by the addition of soil. Artifacts are common. Common in parts of western Europe.

Diagnostic subsurface horizons. Mineral horizons that form below the surface, although they may be exposed by erosion or other processes.

1. **Argillic** - an illuvial horizon (a zone of gain) in which the illuvial material is silicate clay minerals. Since these horizons form due to the movement of clay minerals, they often have coatings of clays (clay skins) on ped faces and pore surfaces.

    a. If any part of the eluvial horizon (zone of loss) has less than 15% clay the argillic must contain at least 3% more clay.

    b. If the eluvial horizon has between 15% and 40% clay there must be 1.2 times more clay in the illuvial horizon.

    c. If the eluvial horizon has more than 40% clay, the illuvial horizon must contain at least 8% more clay.

    The argillic horizon must be at least 1/10 the thickness of the overlying horizons or more than 15 cm thick in sands and loamy sands with lamellas (bands of clay material).

2. **Natric** - Natric horizons are special types of argillic horizons. In addition to the properties of the argillic horizon has more than 15% exchangeable sodium. The exchangeable sodium affects the pH and physical properties of the horizons.

3. **Cambic** - Cambic horizons are illuvial horizons in which there has not been enough clay movement to qualify for an argillic and/or where there is evidence of removal of carbonates.

4. **Spodic** - Spodic horizons are illuvial horizons in which the illuvial material is amorphous materials composed of humus and/or iron and/or aluminum hydrous oxides.

5. **Albic** - A strongly developed eluvial (E) horizon from which clay and free iron oxides have been removed. generally lighter in color than the horizons above or below the albic horizon.

6. **Fragipan** - A loamy subsurface zone, often underlying a B horizon. It is low in organic matter, has high bulk density and is hard to very hard and brittle when dry.

7. **Oxic** - An residual accumulation of hydrated oxides of iron and aluminum or both.

Orders, Suborders, Greatgroups and Subgroups.
Soils are classified into orders and lower levels of classification based on the presence or absence of diagnostic epipedons and subsurface horizons.

A. **Orders.**

1. **Entisols**. Soils that have no subsurface diagnostic horizons nor a mollic epipedon. The lack of B horizon development indicates that these are young soils or soils forming on resistant parent materials.

2. **Inceptisols.** Soils with a cambic B horizon and an umbric, ochric or a plaggen but not a mollic epipedon. These soils are either young soils with slightly more profile development than the entisols or very old soils in which the B horizon has been destroyed by weathering back to a weakly developed cambic B horizon.

3. **Mollisols.** Soils which have a mollic epipedon and at least 50% base saturation if a cambic or an argillic horizon is present. Soils which have dark colored argillic or natric horizons such that when mixed with the surface horizon the mixed zone meets the requirements of a mollic epipedon are also included.

4. **Alfisol**. Soils that do not have a mollic epipedon but that have an argillic horizon with greater than 35% base saturation. If a fragipan is present it must underlie an argillic horizon or meet the requirements of an argillic horizon.

5. **Spodosol**. Soils that have a spodic horizon with an upper boundary within 2 m of the surface or that have a placic horizon (iron cemented pan) that meets all the requirements of spodic horizon except thickness and that rests on a fragipan or spodic horizon.

6. **Ultisols**. Soils with an ochric or umbric epipedon in a mesic or warmer temperature regime that have an argillic horizon or a fragipan that meets all the requirements of an argillic. The base saturation of the argillic or fragipan is less than 35%.

7. **Oxisols**. Soils that have an oxic horizon that is not overlayered by a plaggen epipedon, argillic or natric horizon.

8. **Histisols**. Soils with either 1) more than half of the upper 80 cm composed of organic matter or 2) organic material of any thickness resting on rock or fractured material with the cracks and interstices filled with organic matter.

9. **Vertisols**. Soils which contain, after mixing of the surface 18 cm, 30% or more swelling type clay in all subhorizons down to 50 cm or more, have no lithic or paralithic contact within 50 cm and have at some time cracks open at the surface. Cracks must be at least 1 cm wide at a depth of 50 cm.

10 **Aridisols**. Soils that have an ochric or anthropic epipedon. Theymay or may not have an argillic or natric horizon. If an argillic or natric horizon is present either a salic, petrocalic, calcic, gypsic, petrogypsic or durapan must be present within 1 m of the surface. Soils have an aridic moisture regime.

11 **Andisols**. Soils forming from volcanic rocks, but some may have formed from other pyroclastic rocks. Soils contain less than 25% organic matter and have *andic soil properties*: including high levels of acid oxalate extractable aluminum and iron and very high phosphorus retention capacities.

B. **Suborders**.

Suborder names are formed from the order names by taking a portion of the order name and adding a formative element to the portion to form the order name. Suborder names in the absence of certain major soil properties such as drainage show the macroclimite of the soils. For example consider a Mollisol occurring in a humid climate that is not a poorly or very poorly drained soil.

M_oll_isol h_u_mi_d_ = Udoll

The following suborders are for the orders Mollisols and Alfisols. Suborders for the other 9 orders are formed in similar fashion either showing macroclimate or a major soil property.

1. **Mollisols**.

   a. **Albolls**. Presence of an albic (strongly developed E) horizon.

   b. **Aquolls**. Poorly or very poorly drained mollisols.

   c. **Rendolls**. Mollisols which have no argillic nor calcic, but do have material including coarse fragments 7.5 cm in diameter that have 40% or more calcium carbonate equivalent in or immediately below any mollic epipedon.

   d. **Borolls**. Soils forming or that have formed in a cold climate.

   e. **Xerolls**. Soils forming in a warm climate with an annual dry season.

   f. **Ustolls**. Soils forming in an ustic moisture regime.

   g. **Udolls**. Soils forming in a humid climate.

2. **Alfisols**.

   a. **Aqualfs**. Soils that are somewhat poorly to very poorly drained.

   b. **Boralfs**. Soils forming in a cold climate.

   c. **Ustalfs**. Soils forming in a dry (ustic) climate.

   d. **Xeralfs**. Soils in a warm climate with an annual dry season.

   f. **Udalfs**. Soils in a humid climate.

Laboratory directions

1. The laboratory instructor will provide information (maps etc.) about the field trip.

2. Complete a profile description sheet for each soil pit. Use the descriptive material from laboratories 3 and 4 to aid your profile description.

Note. In case of rain meet in the soil laboratory and do the laboratory on Soil Survey Manuals. The field trip will be rescheduled.

Introductory Soils

Profile Description Sheet

Profile no.____

Horizon	depth	texture	structure	color	mottles (Y/N)
____	______	____	_____	___	___
____	______	____	_____	___	___
____	______	____	_____	___	___
____	______	____	_____	___	___
____	______	____	_____	___	___
____	______	____	_____	___	___
____	______	____	_____	___	___

Parent Material _______________.

Natural Drainage Class ____________________.

Site Position ___________________.

Order ________________.

Suborder _________________.

Great Group ___________________.

Sub Group _________________.

Series _____________________.

Profile no.____

Horizon	depth	texture	structure	color	mottles (Y/N)
____	______	____	_____	___	___
____	______	____	_____	___	___
____	______	____	_____	___	___
____	______	____	_____	___	___
____	______	____	_____	___	___
____	______	____	_____	___	___
____	______	____	_____	___	___

Parent Material ________________.

Natural Drainage Class ____________________.

Site Position ___________________.

Order _________________.

Suborder __________________.

Great Group ___________________.

Sub Group _________________.

Series _____________________.

Profile no.____

Horizon	depth	texture	structure	color	mottles (Y/N)
____	______	____	_____	___	___
____	______	____	_____	___	___
____	______	____	_____	___	___
____	______	____	_____	___	___
____	______	____	_____	___	___
____	______	____	_____	___	___
____	______	____	_____	___	___

Parent Material _______________.

Natural Drainage Class ___________________.

Site Position __________________.

Order ________________.

Suborder ________________.

Great Group __________________.

Sub Group _________________.

Series ___________________.

LABORATORY 5.

USE OF SOIL SURVEY MAPS AND REPORTS.

Soils are distributed in sufficiently uniform patterns and exist as sufficiently recognizable entities to allow mapping and compilation of these maps along with the properties and suggested uses of the soils into reports.

Soil surveys are made cooperatively by federal and state governments. The Soil Conservation Service is the federal agency which has the primary responsibility for soil survey activities. In some surveys, the Forest Service, Bureau of Indian Affairs, Bureau of Reclamation, or Bureau of Land Management may cooperate. Land grant universities and colleges (usually state experiment stations) bear the primary responsibilities in representing the States. Other State agencies such as the Department of Natural Resources or Conservation, and county or city governments may also cooperate. This joint effort is called the **National Cooperative Soil Survey**. The **Standard Soil Survey** is the major type of survey now being produced. The Standard Soil Survey is usually made on a county basis with the counties sharing in the cost of the survey. Less detailed surveys include reconnaissance and soil conservation surveys.

The surveys of soil resources are designed to meet the particular needs of each geographical area. Humid region farmers have different needs than arid region farmers and ranchers that depend on irrigation. Engineers, Foresters, and Conservationists all need different information about soils in a region. The Soil Survey is designed to provide a variety of information about the soils so that all potential uses of the soil can be explored with a high degree of rigor. The Soil Survey is often used to provide equitable tax assessment based on potential soil productivity.

5.1 RECTANGULAR SURVEY.

This system describes the location of a parcel of land relative to the intersection of an east-west **base line** with a north-south **principal meridian**. There are 34 surveyed principal meridians in the U.S.

Lines parallel to the base line are surveyed every six miles north and south of the base line. The six mile wide strips of land are called townships and are assigned a number and direction to identify their distance from the base line (i.e., T 21 N is the township 21 strips north of the base line). Lines parallel to the principal meridian are surveyed every six miles east and west of the principal meridian. These six mile strips of land are called ranges and are also assigned numbers to identify their location relative to the principal meridian (i.e., R6W is the range 6 strips west of the principal meridian). A diagram showing the concept of townships and ranges is Figure 1. Figure 2 shows the base lines and principal meridians used in Illinois.

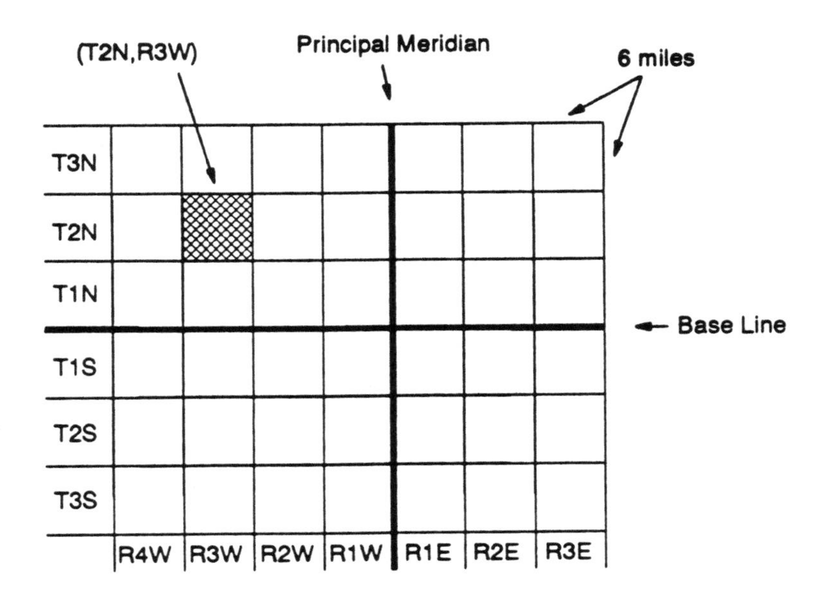

Figure 1. Diagram showing concept of township and range lines.

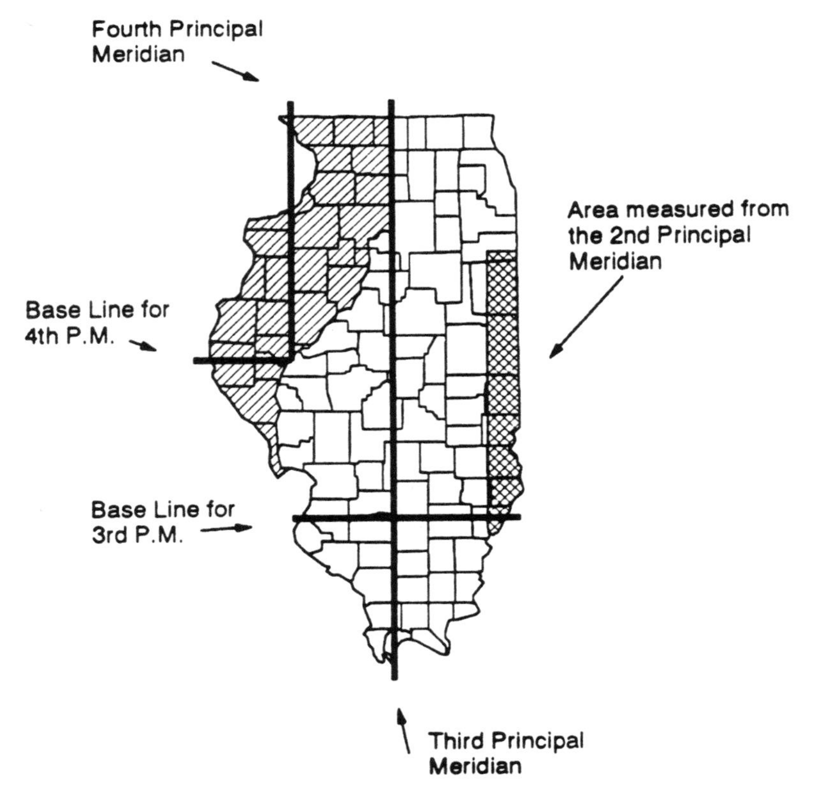

Figure 2. State of Illinois with principal meridians and base lines.

NW 1/4 of NW 1/4 sec 13 17 1

in columbus east township

The intersection of the township strips with the range strips forms squares, six miles on each side (36 square miles). These 36 square mile areas of land are called congressional townships. Townships are further divided into 1 mile square areas of land called sections. Each township contains 36 sections which are numbered starting in the northeast corner. Each section (640 acres) can be further divided into smaller units. Figure 3 illustrates the numbering system used for sections and the how sections are subdivided into smaller units.

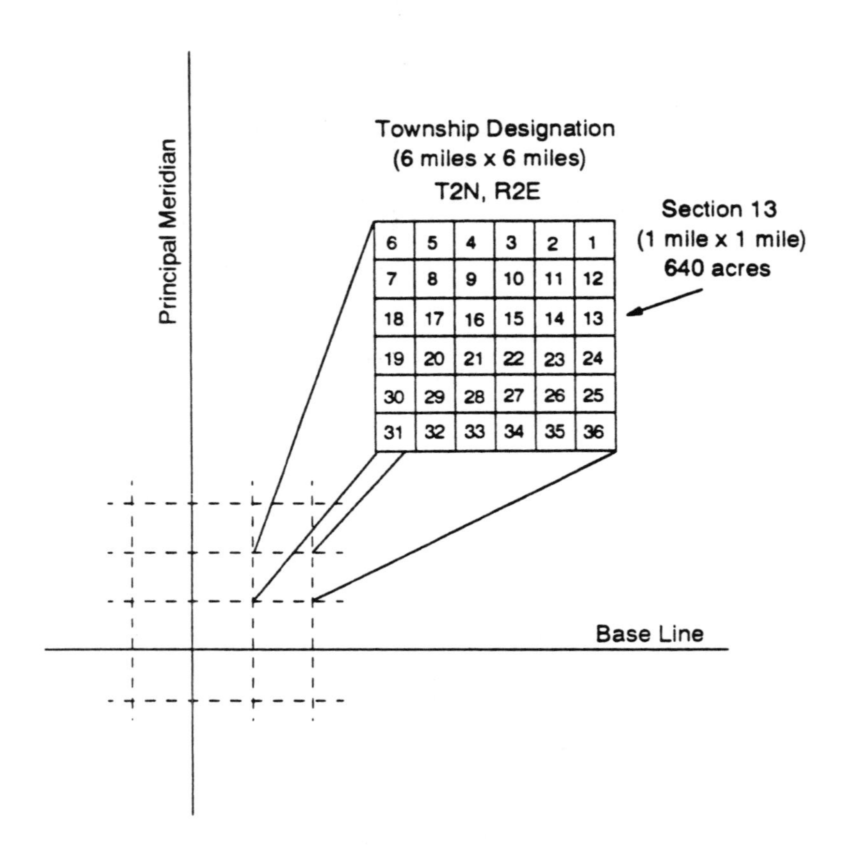

Figure 3. Numbering system and subdivision of sections.

Theoretically, a township is a square tract of land with sides of six miles each, and containing 36 sections of land. In actuality the rectangular survey system imposes a rectangular system on a curved surface. This result in a need to correct for the convergence of the north-south lines in order to keep the east-west parallel.

On the average, section are about 50 feet shorter on the north edge than on the south edge so that every 24 miles (4 townships) a correction line is established parallel to the base line and the sections are again one mile on a side. The deviations from the square that result from the earth's curvature are compensated for in a systematic manner. Sections 1 through 6 on the north side and sections 6, 7, 18, 19, 30 and 31 on the west side may be smaller or larger than 640 acres as required. All other sections are exactly 640 acres whenever possible.

5.2 USE OF SOIL SURVEY REPORTS.

Soil Survey reports are divided into two sections; the first section contains, descriptive material about the county, profile descriptions, soil classifications, tables giving information about engineering uses, soil productivity, wildlife and forest plantings and other uses of individual soils. The section contains detailed soils maps for the county. The maps are drawn on aerial photographs and hence, provide a highly accurate and contain a great deal of detail.

Laboratory procedure.

Answer the following questions using the Peoria County Soil Survey report. The answers are due at the end of the laboratory period. If a different soil survey report is used the instructor will hand out a modified set of questions that are specific for that country.

1. What is the land area of Peoria County?

2. In 1982 how many acres of farm land and woodland were found in Peoria County?

3. What is the average daily maximum temperature in July? What is the average daily minimum temperature in January?

4. After what date would there be only a 10% chance of a 28° F freeze. What is a growing degree day? How many growing degree days are there in Peoria County on the average?

5. In 1982 how many acres of corn, soybeans, wheat and pasture were grown in Peoria County?

6. Identify the parent materials of the following soils: Fayette, Hickory, Marseilles, Dorchester.

7. Identify the pedogenic horizons found in a Catlin soil. Where is the typical pedon of Catlin silt loam located in Peoria County, on what soil map?

8. How many acres of the following soils are found in Peoria County?

   Sawmill, Ipava, Rozetta (all slopes)

9. Is Hickory considered prime farmland? Identify two soils found in Peoria County that are considered to be prime farmland.

10. What is the expected yield of corn and soybeans for the Catlin, Tama and Ipava soils?

11. What are the common trees found on an uneroded Fayette soil? What is the site index for white oak on a Fayette soil?

12. How would Catline and Harpster compare for recreational development?

13. What at the USDA subgroup and family classifications of the following soils?

    Tama, Dorchester, Keomah

14. What are the major soils around the cemetary northwest of Princeville?

15. List the soil numbers of all the soils found in the square mile of land, section 31 of R6E, T11N.

LABORATORY 6.

SOIL WATER AND PLANT-AVAILABLE WATER.

Plants and other organisms are composed primarily of water. Most actively growing plants contain between 75 and 90% water. In addition to the water in the plant cells large quantities of water are lost into the atmosphere through open stomates as part of the transpiration stream. Transpiration occurs because the stomates must be open to facilitate the gas exchange required for photosynthesis. Plants growing on well watered soils lose large amounts of water to the atmosphere. Plants growing on dry soils control water loss by wilting. Wilting prevents the desiccation and death of the plant, but also prevents the exchange of gases necessary for photosynthesis and generally results in lower crop yields.

The following table indicates the approximate amount of water required to produce one pound of above-ground dry matter (stems, leaves and seeds) in a humid region:

Crop	lb. Water/lb. Dry Matter
Sorghum	271
Corn	372
Wheat	505
Barley	521
Cotton	562
Oats	635
Alfalfa	858

Source: Modern Crop Production, Aldrich, Scott and Leng.

Using the above data a 4 ton/acre crop of alfalfa requires approximately 6,800,000 lbs of water. The source of all of this water is water stored in the soil between matric potentials of 1/3 and 15 bars. Soil is the reservoir for water storage and through periodic recharging (rainfall or irrigation) soil is generally capable of supplying the needs of growing plants. Even so, water is usually the most critical factor controlling the growth of terrestrial plants.

6.1 STORAGE OF WATER IN SOILS.

Water is held in soil against the pull of gravity by adhesive and cohesive forces. The adhesive forces are the result of water being chemically bonded to exchangeable cations associated with soil colloids and bonded to polar groups associated with soil minerals and soil organic matter (humus). The cohesive forces are due to hydrogen bonding of additional water molecules to the chemically bonded water resulting in films of water being attracted to soil solids.

When the soil is saturated, that is when all the pores are filled with water, there is both free water and bonded water in the soil. After the soil has drained in response to the pull of gravity only water that is held with some degree of bonding is retained in the soil. Plants can only utilize water that is not strongly held to the soil solids, hence only a portion of the water stored in the soil is available to plants.

If a small amount of water is added to a dry soil, that water will be strongly bonded to the soil solids. If additional water is added to the soil the bonding energy will be spread out over a greater mass of water and the water will be less strongly held to the soil solids. This leads to an important concept about soil water, that is as the amount of water in the soil increases the energy with which the water is held decreases, and conversely as the amount of water in the soil decreases the strength of bonding of water to the soil solids increases.

There is a continuum of energies involved. From very strong bonding (high energies) when only a single molecular layer of water is bonded to the soil solids, decreasing as more and more water is added to the soil down to the point that free water exists in the soil that is not bonded to the soil solids.

Classification of soil water.

Although there is a continuum of energies involved in the bonding of water in soils, there are certain energies that correspond to observable phenomena. The energy of bonding of water to the soil solids is usually expressed as the atmospheres (atm) or bars. This corresponds to the energy required to remove water from the soil.

Saturation The point where all the soil pores are completely filled with water. Water films associated with soil particles are at maximum thickness. Both bonded and free water exist. -- 0 atm matric potential.

Field capacity The point where water ceases to drain out of the soil in response to gravity. Represents the maximum amount of water than can be stored in a soil. -- approximately -1/3 atm matric potential.

Permanent wilting coefficient The point where all water that is energetically available to the plant has been removed. For most plants this corresponds to a matric potential of -15 atm.

Hygroscopic coefficient The water content corresponding to air-dryness, the point where soil water is in equilibrium with atmospheric moisture. Although this point is affected by the relative humidity of the atmosphere is corresponds to approximately a matric potential of -31 atms.

Oven-dry The point where all cohesively bonded water has been removed only chemically (adhesively) bonded water remains -- -10,000 atm matric potential.

Water held between these observable points (potentials) can also be classified.

Free or Gravitational water Nonbonded water that drains in response to gravity.

Capillary water Water held between field capacity and the hygroscopic coefficient, between matric potentials of -1/3 and -31 atms. Functions as the soil solution.

Plant available water Water held between field capacity and the wilting point. Water held at low enough matric potentials that it can be utilized by plants and other soil organisms.

Hygroscopic coefficient Water held in the soil between the hygroscopic coefficient and oven dryness. Water occurring as thin tightly bonded films of water on solid surfaces.

Plant unavailable water Water held at matric potentials that are too large for plants to overcome. Corresponds to matric potentials of -15 atms for most higher plants. Includes a portion of capillary water and hygroscopic water.

Calculation of soil water content.

The amount of water in a soil is most commonly expressed as a percentage of the oven-dry weight of the soil (W%). Unless otherwise stated, water contents in the soils literature and in this class are always expressed by this method.

W% = (wet weight of soil - OD wt. of soil)/OD wt. of soil

W% = weight of water/OD wt. of soil

Example: Calculate the percent water (W%) of a soil that has a wet weight of 120 grams and after drying for 48 hrs. at 110° C has an oven-dry weight (OD wt) of 100 grams.

W% = (120-100)/100 x 100 = 20/100 x 100 = 20%

The reason that the percent of water in soils, as well as the analysis of other materials in soils are expressed on OD wt basis is that this allow consistent comparisons between different soils at different water contents. If analysis were expressed on a wet weight basis the percentage of soil components would decrease after a rain and increase upon drying.

Note: OD wt = Wet wt./1.20 when the soil contains 20% water.

6.2 WATER CONTENT-MATRIC POTENTIAL CURVES.

If a wet soil is placed on a porous plate in a pressure chamber, the pressure can be increased to certain values and the amount of water removed from the soil at that pressure can be determined.

A plot of the logarithm of the matric potential as a function of the water content (W%) produces a water content-matric potential curve.

A water content-matric potential curve allows the calculation of the amount of water in a soil at any matric potential. The amount of plant-available water can be determined for the three soils in Figure 2 by determining the amount of water between matric potentials of -1/3 and -15 atms, that is the difference of the water contents at those two matric potentials.

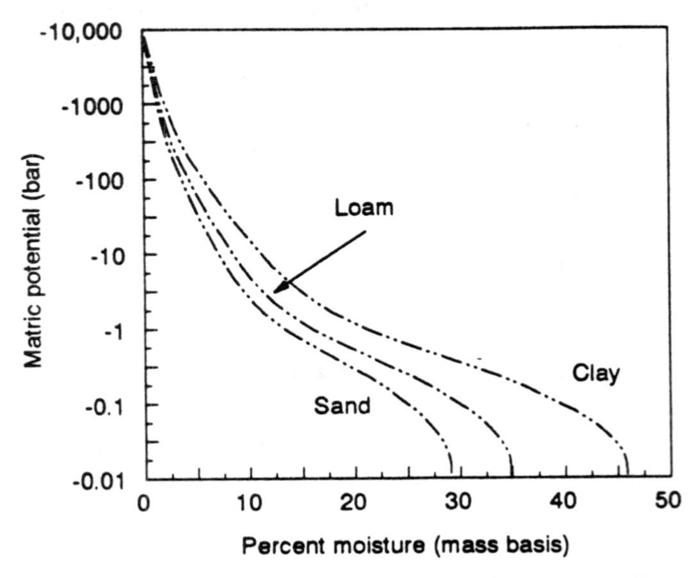

Figure 1 Water content-Matric potential curves for three different textured soils.

Water content-matric potential curves also allow the calculation of the percentage of pore space that is filled with water at different matric potentials. When a soil is saturated 100% of the pores are filled. Hence, the amount of water in the soil at a particular matric potential relative to the water content of the soil at saturation allows the percentage of the soil's pore space that is empty to be calculated from the following equation.

$$\%PS\ empty = (W\%\ at\ sat. - W\%\ at\ X)/(W\%\ at\ sat.) \times 100$$

X is the water content at the desired matric potential where the %PS empty is to be calculated.

6.3 CAPILLARY RISE.

When dry soil is placed in contact with a water table water will be attracted to the soil solids, the initial water will be chemically bonded to the soil matrix and subsequent water will be cohesively bonded through hydrogen bonding to the chemically bonded water. This results in water rising into the dry soil from the water table. Water will rise into the dry soil until the gravitational pull on the mass of water supported by the soil balances the attraction of the soil solids for the water. Since the mass of water supported per unit of chemically bonded water is less in small pores than large pores, the height of rise of water is inversely related to the radius of the pores. That is water will rise higher into small pores than large pores. If the pores are large enough they will not contain water except as thin films of water on solid surfaces. If the pores are small enough they will be completely filled with water.

This attraction of water for solid surfaces gives rise to the property of capillary rise. The diameter of the pores can be related to the maximum height of rise in the following formula:

$$h = 2t/dgr$$

Where h = height of rise (cm)
t = surface tension (75 dynes/cm)
d = density of water (g/cm^3)
g = acceleration due to gravity (980 cm/sec^2)
r = max. capillary radius (cm)

Note: A dyne is the unit of force which will produce an acceleration of 1 cm per second of a 1 gram mass ($g.cm/sec^2$).

Canceling the units and assuming the density of water is equal to 1 gm/cm^3 reduces the capillary rise equation to:

$$h(cm) = 0.15/r(cm)$$

If the equation is expressed in terms of millimeters and in terms of the diameter of a pore instead of its radius the formula becomes:

$$h(mm) = 30/d(mm)$$

6.5 WATER FLOW THROUGH UNIFORM PORES.

The diameter of a pore not only affects the height of capillary rise, but also affects the rate at which water passes through a pore. Since soil structure determines not only the amount of pore space but also the distribution between macropores and micropores, the rate of vertical water movement (drainage) in response to gravity is also a function of structure.

The rate of water movement through uniform pores (such as a garden hose) is given by Poiseuille's equation:

$$Q = P\pi R^4/8LZ$$

Where: Q = quantity of water per unit of time (cm^3/sec)
P = pressure differences ($dynes/cm^2$)
π = 3.1416
R = radius (cm)
L = length of tube or pore (cm)
Z = viscosity of liquid ($dyne \cdot sec/cm^2$)

Note: In the demonstration of flow through different sized tubes all the factors are constant except Q and R. This allows the calculation of the ratio of the radius of the pores from measurement of Q for a set period of time.

$$\frac{Q_{largepores}}{Q_{smallpores}} = \frac{R^4_{largepores}}{R^4_{smallpores}}$$

The demonstration illustrates that doubling the size of the irrigation pipe will deliver 16 times as much water ($2^4 = 16$) as the original pipe. This assume all other factors remain constant.

Laboratory procedure.

1. Construction of an apparent moisture tension curve. Your laboratory instructor will provide soil samples to be used in this experiment.

   a. Cover the bottom of a brass core ring with filter paper. Secure with a rubber band. Repeat procedure with a second piece of filter paper and rubber band. This unit is the "ring assembly"

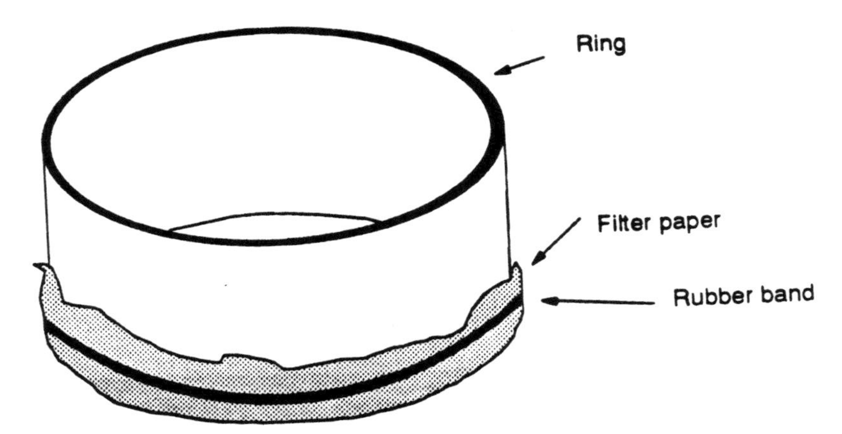

   b. Weigh ring assembly with one and two filter papers and rubber bands, record weights.

   c. Place 50 to 60 grams of air-dry soil into the ring and reweigh the ring assembly and the soil. This weight corrected for the weight of the ring assembly is the air dry (hygroscopic coefficient) weight of the soil.

   d. Place the ring assembly and air dry soil into a pan. Add water to the pan until the water level in the pan is 1 cm <u>below</u> the surface of the soil in the ring assembly. Do not add the water directly to the ring assemblies. These should wet from the bottom up to displace air trapped in the soil pores. Let the ring assembly soak for 10 to 15 minutes.

   e. Fill a moisture can 1/3 full of dry sand and weigh.

   f. Remove the ring assembly from the pan that it was soaking in and place it quickly onto the sand in the moisture can. Weigh the ring assembly and the moisture can. When this weight is corrected for the weight of the ring, can, sand, filter papers and rubber bands it will be the weight of the soil when it is saturated with water.

   g. Allow the ring assembly to drain into the sand in the moisture can for 30 to 60 minutes. Lift the ring assembly out of the moisture can and remove one rubber band and filter paper and then weigh. When this weight is corrected it will be an approximation of the weight of the

17 container 280

soil at field capacity. In the field a soil would require a much longer period to reach field capacity, but this approximation is sufficient for this experiment.

h. Place the ring assembly into a clean weighing can and place in oven (105°C) for 48 hours. After the soil has dried, remove the ring assembly from the oven and allow it to cool in a desiccator. Weigh the ring assembly. After this weight is corrected for the weight of the ring etc. this is the ODwt of the soil. 50 grams soil

Data apparent moisture tension curve. 228.6

Weight of ring and one filter paper and one rubber band (ring assembly 1) 1.3	230
Weight of ring and two filter papers and two rubber band (ring assembly 2) 3.4	232.0
Weight of (ring assembly 2) plus air-dry soil	282
Weight of moisture can plus sand 199	199
Weight of moisture can plus sand and (ring assemble 2) plus saturated soil	529
Weight of (ring assembly 1) plus field-capacity soil	302
Weight of clean dry moisture can (no lid)	X
Weight of moisture can plus ring assembly 1 with OD soil	280

Calculations Apparent moisture tension curve.

1. Wet weight of soil at hygroscopic coefficient (air-dry)(31 bars). 280 229.4 50

   [Weight of ring assembly 2 plus air-dry soil] - [weight of ring assembly 2]

2. Wet weight of soil at saturation (0 bars)

   [Weight of moisture can plus sand and ring assembly 2 plus saturated soil] - [weight of moisture can plus sand and ring assembly 2] = 431

3. Wet weight of soil at field capacity (1/3 bars)

   [Weight of ring assembly 1 plus field capacity soil] - [weight of ring assembly 1]

4. Oven dry weight (10,000 bars)

[weight of ring assembly 1 plus oven-dry soil] - [weight of ring assembly 1] - [weight of clean moisture can]

Using the data from the apparent moisture tension experiment determine the following:

1. Percentage of water at saturation, field capacity, and the hygroscopic coefficient for each soil.

2. Plot a Water content-Matric potential curve (i.e., % water vs. log matric potential) See Figure 1.

3. Estimate percentage of water at permanent wilting point (-15 bars) from the curve.

4. Calculate the percent pore space empty at saturation, field capacity, wilting point, hygroscopic coefficient and oven-dryness.

3. Water flow through uniform pores.

Record the mL of water delivered in a fixed time through the large and small capillary tubes. Using equations from the laboratory manual calculate the ratio of pore diameters for the two capillary tubes.

LABORATORY 7.

SOIL EROSION BY WATER (UNIVERSAL SOIL LOSS EQUATION).

Soil erosion in the midwest is a serious problem. Soil erosion removes the most productive portion of the soil profile and depending upon the quality of the subsoil can result in long term reductions in productivity. In addition, runoff from agriculture land is an important nonpoint source of water pollution. Storm runoff from agriculture land can carry oxygen-demanding organic matter, soil particles and minerals, fecal coliform bacteria, pesticides, fertilizers and other pollutants into surface waters.

The key to preventing the degradation of soil productivity, as well as downstream effects, such as water pollution and siltation of ponds, lakes and reservoirs is the control of erosion. When erosion is controlled to acceptable levels the introduction of pesticides, and fertilizers, such as nitrates, into surface waters are greatly reduced. This is because these materials tend to move bound to soil particles not dissolved in the runoff water. If the rate of flow of water across the soil can be decreased, then the water cannot carry large loads of soil particles and associated pollutants. In addition the lower rates of water flow across the soil surface allow for greater infiltration of the water into the soil and hence, less loss of valuable topsoil.

Sheet and rill erosion occurring on agricultural land is by far the largest source of sediment in Illinois, primarily because about 90% of the state's land is in farms. Urban lands and land where mining or oil extraction is taking place may suffer from very severe erosion, with a great impact on local streams, but their contribution to total erosion is much less than that from farmland. Gully and stream-bank erosion is minor in relation to total soil erosion. However, sediments from these sources can cause severe problems at specific locations.

On agricultural land, most sheet and rill erosion takes place on the gently sloping land (2 to 4% slope). While the rate of erosion per acre is potentially greater on steeper land, Illinois has many more acres of gently sloping cropland. The combined erosion from the gently sloping land generates more total erosion than do the fewer acres of steeper land.

Based on the <u>1970 Conservation Needs Inventory</u>, Illinois has soil losses exceeding four tons per acre per year on 9.7 million acres of cropland, 0.7 million acres of pastureland, 0.6 million acres of woodland, and 0.3 million acres of other land. Soil losses exceeding 20 tons per acre per year occur on an estimated 1.7 million acres of cropland. The acceptable level of erosion from a soil productivity point of view ranges from 1 to 5 tons/A per year (150 tons/A soil loss is approximately equivalent to a loss of one inch of top soil).

The are approximately 11 million acres of Illinois cropland where erosion is at or below an accepted soil loss level. The other approximately 13 million acres of cropland are being farmed in such a manner as to produce unacceptable levels of soil erosion.

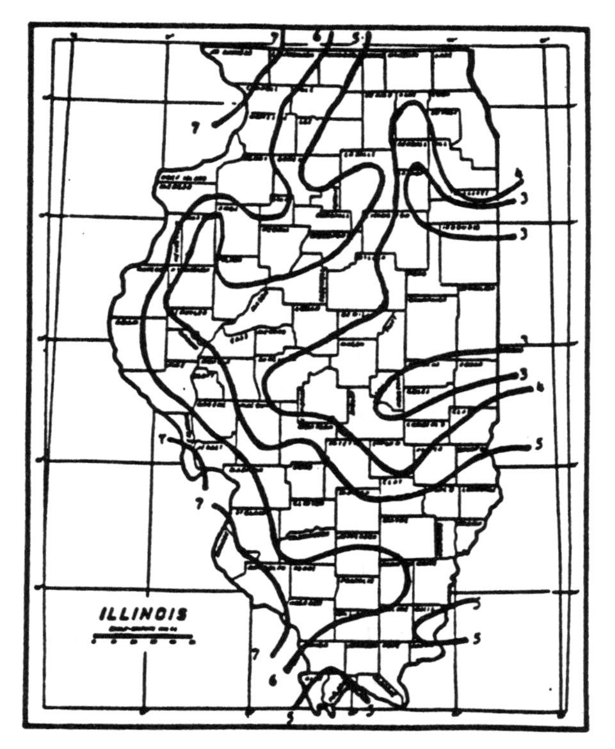

Figure 1 Average annual soil erosion (sheet and rill) rates (tons/acre)

7.1 PREDICTING SOIL LOSS WITH THE UNIVERSAL SOIL LOSS EQUATION.

The Universal Soil Loss Equation provides a method for determining the various combinations of conservation cropping systems and mechanical practices which under the type and distribution of rainstorms expected in a particular locality will result in satisfactory erosion control on each specific field. Evaluation of a combination of conservation practices is easily done and predicted soil loss quickly determined.

The Universal Soil Loss Equation is: $A = R\,K\,LS\,C\,P$

where;

A is the computed soil loss per unit area (tons of soil loss per acre).

R is the rainfall and runoff factor.

K is the soil erodibility factor. It is the soil loss rate for a given soil type as determined from experimental plots.

LS are the soil-length and soil-steepness factors.

C is the cover and management factor, it is the ratio of soil loss from an area of specified cover and management to that from an identical area in tilled continuous fallow.

P is the conservation practice factor, it is the ratio of soil loss with a conservation practice to straight row farming up and down the slope.

An estimate of the rate of soil erosion that will be caused by water is obtained by looking up values of R, K, LS, C, and P in appropriate maps and tables and multiplying them together to obtain the computed soil loss per unit area (A). This estimate is the compared to the rate of soil loss (T) that can be tolerated agronomically by a given soil type. The data used in compilation of the Universal Soil Loss Equation does not measure erosion losses from snow-melt; only rainfall is considered. However, in certain areas of the U.S. a sub-factor for thaw and snowmelt can be used.

7.1.1 R - rainfall factor.

Basic data from Weather Bureau Stations covering a period of 22 years were used in preparation of R factors. Soil loss from cultivated fallow land has been found to be directly proportional to the product of two rainfall characteristics: kinetic energy (E) of the rain times the maximum 30 minute intensity (I) of the rain. The sum of these products, called EI values provides a numerical rainfall erosion index that evaluates the erosion potential of the rainfall within a year. The average annual total EI values for a particular locality is the rainfall erosion index (R) for that locality and are obtained from Figure 2. R values vary from about 160 in Northern Illinois to 250 in Southern Illinois.

Figure 2. Rainfall R Factor Map.

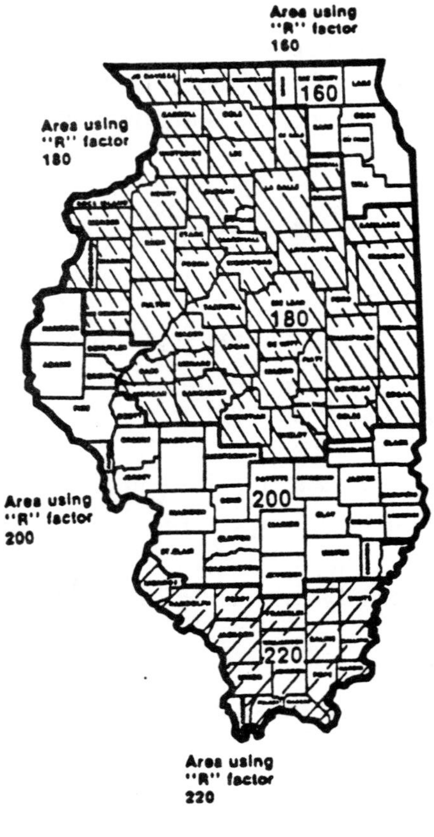

7.1.2 K - soil erodibility factor.

The K factor is a quantitative value the has been experimently determined. For a particular soil, it is the rate of soil loss as measured on a standard plot. A soil's erodibility is a function of complex interactions between its chemical and in particular the soil's physical properties. Some the more important physical properties are structure, texture, organic matter content and soil depth. In Illinois a soil type becomes less erodible with a decrease in silt fraction, regardless of whether the corresponding increase is in the sand or the clay fraction.

Soil erodibility by water is influenced by (a) soil properties that affect infiltration, permeability, and water holding capacity, and (b) soil properties the influence the dispersion, splashing, abrasion and transporting forces of rainfall and runoff. Soil erodibility factors (K factors) based on general soil properties, with examples of soil series are given in Table 1.

Table 1. K values for certain soil types.

Soil properties	K values	T Values[a]
Dark- and moderately dark-colored soils, somewhat poorly and poorly drained with good permeability (Muscatune, Ipava, Flanagan, Sable, Drummer and Herrick)	0.28	5
Dark- and moderately dark-colored soils, well to moderately well drained with good permeability (Catlin, Harrison, Proctor, Saybrook, and Tama)	0.32	5 - 4
Dark- and light-colored soils with restricted permeabilities (Cisne, Cowden, Clarance)	0.37	3 - 2
Dark-colored soils with very restricted permeabilities (Swygert)	0.43	3 - 2
Light-colored soils with good permeabilities (Alford, Birkbeck, Clinton, and Fayette)	0.37	5 - 4
Light-colored soils with restricted permeabilities (Ava, Blount, Grantsburg, Hosmer, and Wynoose)	0.43	4 - 3
Sandy loam soils (Dickinson, Onarga, and Ridgeville)	0.20	4 - 3
Loose sands (Ade, Plainfield, and Sparta)	0.17	5

a. The first figure represents the soil-loss tolerance for soils with less than severe soil erosion. The second figure represents the soil-loss tolerance for soil with severe erosion and strong evidence of subsoil mixing with the topsoil.

LS - slope length and steepness factor.

Both the length of slope (L) and the steepness of the slope (S) substantially affect the rate of soil erosion by water. The two effects have been evaluated separately in research and are represented in the soil loss equation by LS. To determine the value of LS use Table 2, first find the % slope and then read across to the length.

Both slope length and slope steepness must be determined on the specific field segment where you are estimating soil erosion. Slope length is the distance from the point of origin of runoff to the point where (1) the slope gradient decreases enough that sediment deposition begins, (2) the runoff water becomes a concentrated flow, or (3) the runoff enters a well-defined channel that may be part of the natural drainage network or a constructed channel such as a grass waterway or terrace channel.

Table 2. LS values for specific combinations of slope length and steepness.

Slope	25 ft	50 ft	75 ft	100 ft	150 ft	200 ft	300 ft	400 ft	500 ft	600 ft	800 ft	1000
0.2%	0.06	0.07	0.08	0.08	0.09	0.09	0.10	0.11	0.11	0.11	0.12	0.12
0.5%	0.07	0.08	0.09	0.10	0.10	0.11	0.12	0.13	0.13	0.14	0.15	0.15
0.8%	0.09	0.10	0.11	0.11	0.12	0.13	0.14	0.15	0.16	0.16	0.17	0.18
2%	0.13	0.16	0.19	0.20	0.23	0.25	0.28	0.31	0.33	0.34	0.38	0.40
3%	0.19	0.23	0.26	0.29	0.33	0.35	0.40	0.44	0.47	0.49	0.54	0.57
4%	0.23	0.30	0.36	0.40	0.47	0.53	0.62	0.70	0.76	0.82	0.92	1.01
5%	0.27	0.38	0.46	0.54	0.66	0.76	0.93	1.07	1.20	1.31	1.52	1.69
6%	0.34	0.48	0.58	0.67	0.82	0.95	1.17	1.35	1.50	1.65	1.90	2.13
8%	0.50	0.70	0.86	0.99	1.21	1.41	1.72	1.98	2.22	2.43	2.81	3.14
10%	0.69	0.97	1.19	1.37	1.68	1.94	2.37	2.74	3.06	3.36	3.87	4.33
12%	0.90	1.28	1.56	1.80	2.21	2.55	3.13	3.61	4.04	4.42	5.11	5.71
14%	1.15	1.62	1.99	2.30	2.81	3.25	3.98	4.59	5.13	5.62	6.49	7.26
16%	1.42	2.01	2.46	2.84	3.48	4.01	4.92	5.68	6.35	6.95	8.03	8.98
18%	1.72	2.43	2.97	3.43	4.21	4.86	5.95	6.87	7.68	8.41	9.71	10.9
20%	2.04	2.88	3.53	4.08	5.00	5.77	7.07	8.16	9.12	10.0	11.5	12.9

Terracing shortens the length of slope. You can evaluate the effectiveness of terracing on a specific field by substituting terrace spacing for slope length in the equation. The most common terrace spacings used in Illinois range from 120 to 200 feet.

Use of the first three factors.

R x K x LS represent the estimated amount of soil that would be lost from a particular field segment if it were maintained in fallow condition the entire year (cultivated by no crop or other cover grown). If no crop is grown (C) nor any conservation practice (P) implemented, then each of these factors in the Universal Soil Loss Equation has a value of 1. However, if either factor is implemented, i.e., a crop grown or conservation practice used the value of the C and P factors becomes less than 1 representing a reduction in soil loss from year round fallow.

C - cover and management factor.

The values for C have been developed from research dealing with cropping systems and are found in Tables 3, 4, and 5. Each table represents a different portion of the state. C represents the reduction in soil erosion when a specific cropping system is compared with continuous fallow (soils tilled but no crop grown). A unique C value a given cropping system has been developed for each section of the state reflecting factors such as different climatic conditions, planting dates. For example, in northern Illinois approximately 80 percent of the year's soil erosion occurs from May 1 to October 1, whereas only 60 percent of the annual soil erosion occurs during the same period in southern Illinois. In northern Illinois the soils freeze in the winter and the precipitation is in the form of snow, whereas in southern Illinois soils are frozen for a much shorter time and more of the precipitation is in the form of rain. Also, in southern Illinois a higher percent of the annual precipitation occurs in the winter. Use Figure 3 to determine which of the three C factor tables to use for your portion of the state.

Figure 3. Crop management C factor map.

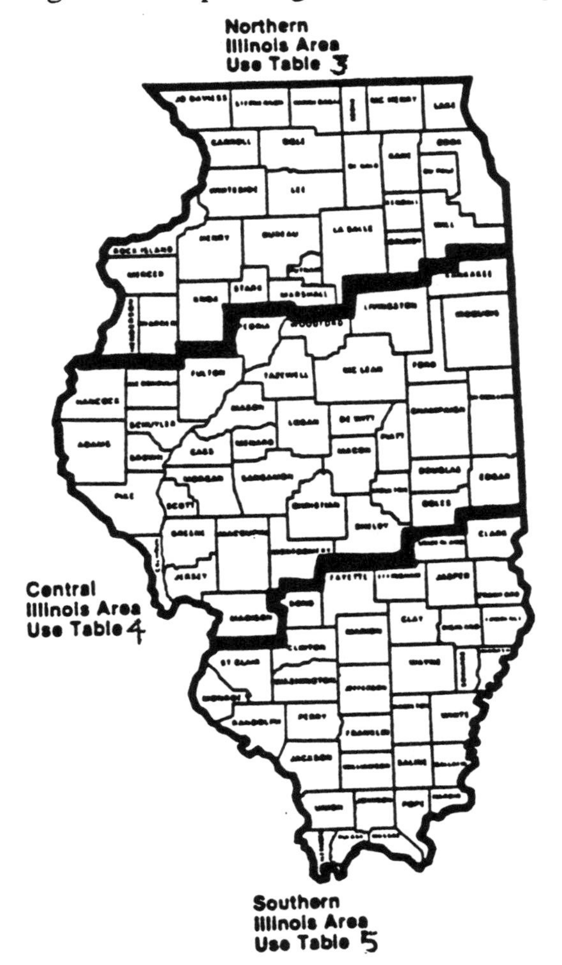

Table 3. Crop management C factors for northern Illinois.

Crop sequence	Conventional-tillage systems				Plow plant or wheel track plant	Conservation-tillage systems									
						Chisel planting, till plant or strip plant					Zero-till, no-till, or slot plant				
	Fall Plow Conventional tillll		Spring plow Conventional till			Corn residue or equivalent on surface					Corn residue or equivalent on surface				
	RDL	RDR	RDL	RDR	RDL	1000 to 2000 lb	2000 to 3000 lb	3000 to 4000 lb	4000 to 6000 lb	over 6000 lb	1000 to 2000 lb	2000 to 3000 lb	3000 to 4000 lb	4000 to 6000 lb	over 6000 lb
Cont. Soybeans	0.51	NA	0.47	NA	0.37	0.33	0.22	NA	NA	NA	0.31	0.17	NA	NA	NA
C-Sb	0.45	0.51	0.42	0.47	0.30	0.32	0.21	0.16	0.12	0.098	0.30	0.17	0.12	0.065	0.041
C-C-Sb	0.43	0.48	0.40	0.45	0.28	0.32	0.21	0.16	0.12	0.098	0.30	0.17	0.12	0.065	0.041
Cont. Corn	0.40	0.48	0.37	0.45	0.25	0.32	0.20	0.12	0.072	0.038	0.30	0.17	0.12	0.064	0.030
C-C-Sb-Gx	0.35	0.38	0.32	0.36	0.25	0.27	0.19	0.14	0.097	0.070	0.20	0.13	0.11	0.081	0.063
C-Sb-Gx	0.33	0.35	0.31	0.33	0.25	0.26	0.18	0.14	0.10	0.081	0.16	0.12	0.10	0.087	0.075
C-C-C-Gx	0.31	0.36	0.29	0.34	0.21	0.25	0.16	0.11	0.075	0.050	0.18	0.11	0.086	0.060	0.045
C-C-Sb-G-M	0.22	0.25	0.20	0.24	0.15	0.19	0.15	0.12	0.093	0.080	0.15	0.10	0.082	0.061	0.047
C-Sb-G-M	0.18	0.19	0.17	0.18	0.12	0.16	0.13	0.11	0.099	0.090	0.12	0.086	0.072	0.060	0.051
C-C-G-M	0.14	0.17	0.13	0.16	0.085	0.14	0.11	0.091	0.078	0.069	0.097	0.066	0.052	0.039	0.031
C-G-M	0.073	0.11	0.064	0.098	0.042	NA	NA	NA	NA	NA	NA	NA	NA	0.024	0.021

C = corn, Sb= soybeans, G = small grain, Gx = small grain with catch crop, M = meadow

Table 4. Crop management C factors for central Illinois.

Crop sequence	Conventional-tillage systems				Plow plant or wheel track plant	Conservation-tillage systems									
						Chisel planting, till plant or strip plant					Zero-till, no-till, or slot plant				
						Corn residue or equivalent on surface					Corn residue or equivalent on surface				
	Fall Plow Conventional till		Spring plow Conventional till			1000 to 2000 lb	2000 to 3000 lb	3000 to 4000 lb	4000 to 6000 lb	over 6000 lb	1000 to 2000 lb	2000 to 3000 lb	3000 to 4000 lb	4000 to 6000 lb	over 6000 lb
	RDL	RDR	RDL	RDR	RDL										
Cont. Soybeans	0.55	NA	0.49	NA	0.40	0.33	0.23	NA	NA	NA	0.29	0.16	NA	NA	NA
C-Sb	0.47	0.53	0.43	0.49	0.33	0.32	0.22	0.16	0.12	0.10	0.29	0.16	0.11	0.063	0.040
C-C-Sb	0.44	0.51	0.40	0.47	0.30	0.31	0.21	0.16	0.12	0.10	0.29	0.16	0.11	0.063	0.040
Cont. Corn	0.40	0.51	0.37	0.47	0.26	0.31	0.20	0.12	0.071	0.040	0.29	0.16	0.11	0.061	0.030
C-C-Sb-Gx	0.35	0.40	0.32	0.36	0.25	0.26	0.18	0.13	0.089	0.065	0.18	0.12	0.10	0.073	0.056
C-Sb-Gx	0.33	0.36	0.30	0.33	0.24	0.24	0.17	0.13	0.095	0.073	0.15	0.11	0.092	0.077	0.065
C-C-C-Gx	0.31	0.38	0.28	0.34	0.21	0.24	0.16	0.11	0.071	0.048	0.17	0.11	0.081	0.055	0.040
C-C-Sb-G-M	0.22	0.27	0.20	0.24	0.15	0.18	0.14	0.11	0.088	0.075	0.14	0.094	0.074	0.054	0.041
C-Sb-G-M	0.18	0.20	0.16	0.18	0.12	0.15	0.13	0.11	0.093	0.084	0.11	0.076	0.064	0.063	0.044
C-C-G-M	0.14	0.18	0.12	0.16	0.083	0.13	0.11	0.087	0.075	0.067	0.090	0.060	0.047	0.035	0.027
Double crop Corn-Soybeans	NA	NA	0.36	NA	NA	NA	NA	NA	NA	NA	NA	NA	0.15	0.11	0.089
C-G-M	0.072	0.10	0.041	0.093	0.041	NA	NA	NA	NA	NA	NA	NA	NA	0.021	0.018

C = corn, Sb= soybeans, G = small grain, Gx = small grain with catch crop, M = meadow

Table 5. Crop management factors for southern Illinois.

Crop sequence	Conventional-tillage systems				Plow plant or wheel track plant	Conservation-tillage systems									
						Chisel planting, till plant or strip plant					Zero-till, no-till, or slot plant				
	Fall Plow Conventional till		Spring plow Conventional till			Corn residue or equivalent on surface					Corn residue or equivalent on surface				
	RDL	RDR	RDL	RDR	RDL	1000 to 2000 lb	2000 to 3000 lb	3000 to 4000 lb	4000 to 6000 lb	over 6000 lb	1000 to 2000 lb	2000 to 3000 lb	3000 to 4000 lb	4000 to 6000 lb	over 6000 lb
Cont. Soybeans	0.61	NA	0.51	NA	0.42	0.33	0.26	NA	NA	NA	0.27	0.14	NA	NA	NA
C-Sb	0.51	0.61	0.44	0.54	0.35	0.32	0.25	0.17	0.14	0.12	0.26	0.14	0.10	0058	0.038
C-C-Sb	0.48	0.59	0.40	0.51	0.32	0.30	0.23	0.16	0.13	0.11	0.26	0.14	0.10	0.058	0.038
Cont. Corn	0.42	0.59	0.36	0.53	0.27	0.29	0.21	0.12	0.069	0.046	0.26	0.14	0.10	0.053	0.030
C-C-Sb-Gx	0.34	0.42	0.29	0.36	0.23	0.25	0.19	0.13	0.097	0.081	0.17	0.12	0.10	0.078	0.067
C-Sb-Gx	0.31	0.37	0.26	0.31	0.21	0.23	0.18	0.14	0.11	0.093	0.13	0.11	0.097	0.086	0.079
C-C-Sb-G-M	0.22	0.28	0.18	0.24	0.13	0.18	0.15	0.12	0.096	0.087	0.13	0.090	0.075	0.058	0.049
C-G-M	0.070	NA	0.050	NA	0.032	NA	NA	NA	NA	NA	NA	NA	NA	0.023	0.021
C-Sb-G-M	0.17	0.21	0.14	0.17	0.11	0.16	0.14	0.12	0.10	0.098	0.095	0.076	0.067	0.059	0.054
C-C-M-M	0.14	0.21	0.11	0.17	0.078	NA	NA	NA	NA	NA	0.071	0.045	0.036	0.024	0.081

C = corn, Sb= soybeans, G = small grain, Gx = small grain with catch crop, M = meadow

C factors for permanent pasture, range, idle land as well as undisturbed forest land are given in Tables 6 and 7.

Table 6. C factors for undisturbed forest land.

Area covered by canopy of trees and undergrowth (%)	Area covered by duff at least 2 inches deep (%)	C factor
100 to 75	100 to 90	0.0001 to 0.001
75 to 45	85 to 75	0.002 to 0.004
40 to 20	70 to 40	0.003 to 0.009

The range in listed C factors is the result in differences found in specified forest litter and canopy covers and by variations in effective canopy heights.

Table 7 C factors for permanent pasture, range and idle land[a]

Vegetative canopy		Cover that contacts the soil surface						
Type and height	Percent cover	Ground cover (percent)						
		Type	0	20	40	60	80	95+
No appreciable canopy		G	0.45	0.20	0.10	0.042	0.013	0.003
		W	0.45	0.24	0.15	0.091	0.043	0.011
Tall weeds or short brush with average drop fall height of 20 inches	25	G	0.36	0.17	0.09	0.038	0.013	0.003
		W	0.36	0.20	0.13	0.083	0.041	0.011
	50	G	0.26	0.13	0.07	0.035	0.012	0.003
		W	0.26	0.16	0.11	0.076	0.039	0.011
	75	G	0.17	0.10	0.06	0.032	0.011	0.003
		W	0.17	0.12	0.09	0.068	0.038	0.011
Appreciable brush or bushes, with average drop fall height of 6.5 feet	25	G	0.40	0.18	0.09	0.040	0.013	0.003
		W	0.40	0.22	0.14	0.087	0.042	0.011
	50	G	0.34	0.16	0.08	0.038	0.012	0.003
		W	0.34	0.19	0.13	0.082	0.041	0.011
	75	G	0.28	0.14	0.08	0.036	0.012	0.003
		W	0.28	0.17	0.12	0.078	0.040	0.011
Trees, but no appreciable low brush. Average drop fall height of 13 ft.	25	G	0.42	0.19	0.10	0.041	0.013	0.003
		W	0.42	0.23	0.14	0.089	0.042	0.011
	50	G	0.39	0.18	0.09	0.040	0.013	0.003
		W	0.39	0.21	0.14	0.087	0.042	0.011
	75	G	0.36	0.17	0.09	0.039	0.012	0.003
		W	0.36	0.20	0.13	0.084	0.041	0.011

The listed C values assume that the vegetation and mulch are randomly distributed over the entire area.

Canopy height is measured as the average fall height of water drops falling from the canopy to the ground. Canopy effect is inversely proportional to drop fall height and is negligible if fall height exceeds 33 feet.

Percent cover is the portion of total-area surface that would be hidden from view by canopy in a vertical projection (bird's-eye view).

G Cover at surface is grass, grasslike plants, decaying compacted duff, or litter at least 2 inches deep.
W Cover at surface is mostly broadleaf herbaceous plants or undecayed residues or both.

Conventional tillage includes using the moldboard plow, disking, planting and cultivating. Generally conventional tillage removes most of the crop residue leaving the soil exposed to the harmful effects of rainfall. Conservation tillage may be carried out with either a chisel plow or disk as the primary tillage tool, followed by a field cultivator or other secondary tillage tools that leave a portion of the crop residue on the soil surface. The effectiveness of a conservation tillage system depends primarily on the amount of crop residue left on the soil surface until the new crop becomes established and the crop canopy provides soil-erosion protection.

Tables 3, 4, and 5 include columns showing different amounts of crop residues left on the soil surface after planting. Average crop residues vary with each crop and yield level, but the following residue estimates may be used as a guide. Corn produces residue at approximately 1 pound of residue to each pound of shelled corn produced. A 100 bushel corn crop will produce 5,600 pounds of crop residue. Small grains such as wheat produce about 100 lbs of crop residue for a bushel of grain; a 50 bushel wheat crop will produce 5,000 lbs of crop residue. Soybeans will produce about 45 lbs of crop residue for each bushel of beans. A 50 bushel soybean crop will produce about 2,250 lbs of residue.

Use the numbers in Table 8 to estimate the amount of crop residues that will be left on the soil surface after tillage operations. For example, assume a 125 bushel corn crop producing 7,000 lbs of residue. The field is disked and the fall chiseled. Approximately 60%, or 4,200 lbs of residue will remain on the surface after disking, and approximately 75%, or 3,150 lbs, will remain after chiseling. If winter losses are approximately 30% (a good average) then 2,200 lbs will remain on the soil surface in the spring. If the field is tilled in the spring with a field cultivator prior to planting this will leave about 75% or 1,650 lbs on the surface after planting.

Table 8. Estimates of residue remaining after various tillage operations.

Tillage operations	Approximate percent of residue remaining after one trip (corn residue)
No-till planting	95 to 100
Chisel plow (straight shanks)	75 to 80
Chisel plow (twisted shanks)	40 to 50
Field cultivator (with sweeps)	75 to 80
Tandem disk	40 to 60
Offset disk (24-inch blades, 6 inches deep)	25 to 50
Moldboard plow	0 to 5
Overwinter decomposition	70

Residues remaining will vary depending on field speed, depth of setting, and soil conditions at the time of operation.

Wheat and soybean residues are approximately twice as effective as corn stalks in controlling soil erosion. If you estimate that there is 500 lbs of soybean or wheat residue left after planting, use the 1,000 lbs column in the tables. Zero tillage leaves the soil surface nearly undisturbed and except for the approximately 30% winter loss use a residue amount based only on crop type and yield.

P - conservation practices.

The P factor represents the reduction in soil erosion after the use of conservation practices. If no conservation practice is used the value of P is one. Table 9 gives the P factors for a few different conservation practices; contouring and contour strip cropping with either a row crop-small grain-meadow-meadow or row crop-row crop-small grain-meadow rotation over a four year period.

Table 9. P factors for conservation practices.

Percent slope	Contouring	Contour strip cropping R-G-M-M	Contour strip cropping R-R-G-M
1.1 to 2.0	0.60	0.30	0.45
2.1 to 7.0	0.50	0.25	0.40
7.1 to 12.0	0.60	0.30	0.45
12.1 to 18.0	0.80	0.40	0.60
18.1 to 24.0	0.90	0.45	0.70

R = row crop (corn or seybeans), G = small grain, M = meadow.

Full contouring benefits are obtained only when the field is relatively free from gullies and depressions other than grass waterways. As slope length increases contouring becomes less and less effective. Reasonable of the limits that slope length places on contouring are: 400 feet for 1 to 2% slopes, 300 feet for 3 to 5% slopes, 200 feet for 6 to 8% slopes and 100 feet for 9 to 14% slopes.

A - predicted average annual soil loss.

A represents the average annual soil erosion loss in tons per acre due to water erosion. It is the average annual soil loss for the specific field for which you have determined the appropriate equation factors.

$$A = R \times K \times LS \times C \times P$$

Example.

Assume that you are farming a Clinton silt loam soil with a 5 percent slope 300 feet long in Pike county Illinois. The R value for Pike county is 200 (Figure 2), the K value for Clinton silt loam is 0.37 (Table 1), and LS is 0.93 (Table 2).

Hence: 200 x 0.37 x 0.93 = 68.8 tons of soil lost under fallow.

If the crop rotation is corn, soybean and wheat with a clove catch crop using conventional tillage, spring plowed and residues left on the soil surface, the C factor is 0.30 (Table 4).

68.8 x 0.30 = 20.6 tons soil lost under the rotation.

Because the soil-loss tolerance level (T) for a Clinton silt loam is 5 tons of soil loss per acre per year. Either the cropping system must be altered to produce a smaller C factor or conservation practices must be used to reduce the P factor. These are the two factor that are under the control of the farmer.

If the field is contoured this will change the P factor from 1 to 0.5 reduce the erosion by 50%.

20.6 x 0.5 = 10.3 tons soil loss per acre per year.

Then if the tillage system were changed to conservation tillage, such as chisel plowing, leaving 3,000 to 4,000 pounds of crop residue on the soil surface after planting, the C factor would be reduced from 0.30 to 0.13 further decreasing soil erosion.

200 x 0.37 x 0.93 x 0.13 x 0.5 = 4.5 tons/acre/year.

These changes would bring the annual soil loss below the 5 ton soil erosion loss limit and protect the long term productivity of the soil.

Laboratory procedure.

Your lab instructor will provide you with a set of situations. You will be asked to estimate the annual soil loss and suggest conservation practices or other changes to bring the predicted soil loss into acceptable limits.

LABORATORY 8.

CATION EXCHANGE CAPACITY OF SOILS.

8.1 GENERAL CONCEPTS.

Cation exchange is the reversible, low energy transfer of ions between solid and liquid phases. Cation exchange affects many soil processes:

1. Weathering of soil minerals.
2. Nutrient absorption by plants.
3. Leaching of electrolytes.
4. Buffering of soil pH.

Cation exchange is the result of the neutralization of the negative charge on soil colloids by oppositely charged cations. The cations are held to the colloidal surface by columbic attraction with van der Waal forces and induction forces increasing the strength of bonding of some types of cations.

Along with the attraction and subsequent concentration of cations which are oppositely charged to the colloidal surface, is the repulsion of anions which are negatively charged like the colloidal surface.

8.2 CONCEPTUAL MODEL OF THE CATION EXCHANGE PROCESS.

Cations are not rigidly held on the colloidal surface, but because of their thermal energies have some degree of motion on and away from the surface. This movement around on the surface defines a hemisphere of motion for each particular combination of ion and colloidal surface. Consider two cases, the first case, a tightly held and hence, nonexchangeable cation and the second case, a less tightly held exchangeable cation.

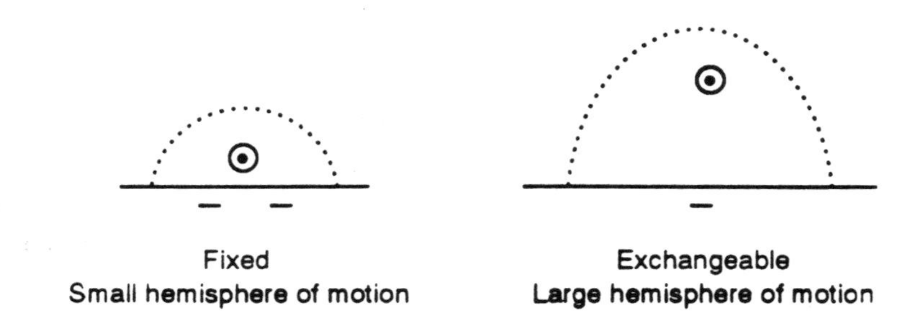

Figure 1 Fixed vs exchangeable cations.

What is cation exchange? Cation exchange occurs (Fig. 2a) when ions (⊗) in the bulk solution move into the hemisphere of motion of a cation on the surface (⊙) at a point in time when the exchangeable cation is far from the surface. The ion initially in solution becomes trapped on the surface by the negative charge, the ion initially on the surface moves into the soil solution. If and ion is close to the charged colloidal surface when the solution ion randomly moves into the hemisphere of motion ion exchange does not occur (Fig. 2b) and the ion returns to the solution. The motion of the ions in the bulk solution is due to their thermal energies.

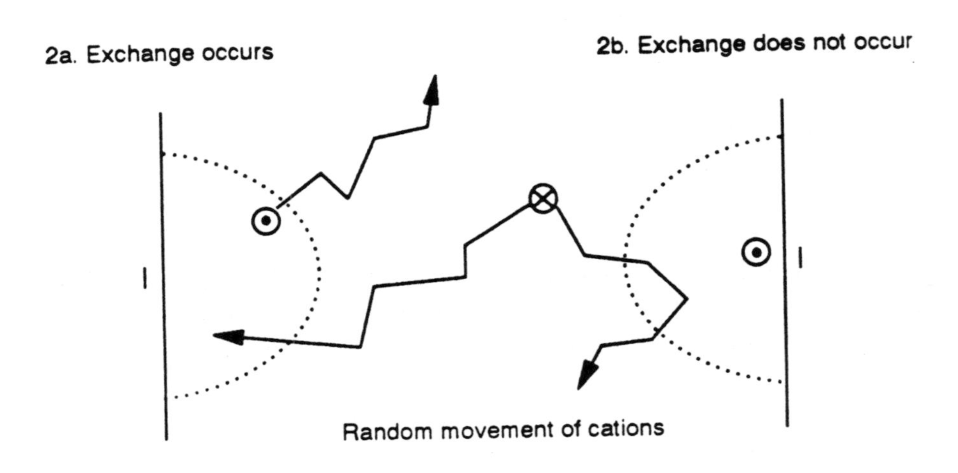

Figure 2a & b Example of cation exchange.

Many factors effect the distribution of cations between the soil solution and the colloidal surface. An intuitive feel for the factors effecting cation exchange can be developed using the conceptual model.

1. As the concentration of a particular cation in the bulk solution increases the probability of that type of cation penetrating the hemisphere of motion of a surface cation at a time when the surface cation is distance from the surface also increases. Hence, as the concentration of an cation in solution increases, there is a corresponding increase in the amount of that cation on the colloidal surface (exchange site).

2. The effect of ion valence can also be illustrated. As the valence of an exchangeable cation increases so does the affinity of the cation for the surface resulting in a smaller hemisphere of motion. Hence, as valence increases the exchangeability of the ion decreases and the cations concentration on the colloidal surface increases relative to cations of lower valence.

Other factors such as the hydrated size of the cation and the density of charge on the colloidal surface also affect the degree of attraction of the cation to the colloidal surface. The hydrated size of the cation determines how close the cation can approach the negative charged surface, while the density of charge on the surface affects the strength of attraction of the cation. Both factors affect the cations hemisphere of motion and therefore, its exchangeability.

8.3 MASS ACTION CONCEPT OF CATION EXCHANGE.

Cation exchange can be treated as a mass action chemical reaction, where cations in the soil solution compete for exchange sites on the soil colloids based on their concentrations in the soil solution.

$$2\text{Na-clay} + Ca^{2+}{}_{aq} = \text{Ca-clay} + 2Na^{+}{}_{aq}$$

When the concentration of $Ca^{2+}{}_{aq}$ in the soil solution is increased the reaction is shifted to the right increasing the concentration of calcium on the clay and releasing Na^{+} ions to the soil solution. If the concentration of calcium ions in the soil solution is decreased then the reaction will shift to the left releasing calcium ions to the soil solution. The above equation illustrates that the ionic make up of the soil solution and the colloids are connected and cannot be varied independently. When the concentration of an cation in the soil solution is increased by processes such as fertilization or decreased by processes such as nutrient absorption by plants, the concentration of the cation on the exchange sites must also increase or decrease.

Analysis of the soil shows that the amount of cations in the soil solution is small compared to the amount of cations held on the colloidal surfaces as exchangeable cations. The actual ratio of total exchangeable cations to total cations in the soil solution depends primarily on the exchange capacity of the soil. The ionic composition of the soil solution is reasonably constant from soil to soil, with the exception of salt-affected soils. Whereas the amount of exchangeable cations is a function of the amount and type of clay and the humus (organic colloid) content of the soil.

8.4 CATION EXCHANGE CAPACITIES (CEC) OF CLAY MINERALS.

8.4.1 1:1 minerals.

Minerals such as kaolinite have little or no isomorphous substitution. Because of the multiple hydrogen bonding between the layers they are nonswelling. The only source of charge is the pH dependent charge sites on crystal edges and at surface irregularities. The low negative charge of coupled with the nonswelling nature of these clays results in minerals with low exchange capacities.

8.4.2 2:1 minerals.

The cation exchange capacity of these minerals is related to the amount of isomorphous substitution, whether the cations balancing the negative charge are tightly or weakly held and by the ability of the clay mineral to swell.

a. <u>Micas</u>. Most if not all the permanent charge is satisfied by "fixed" cations. There is little or no swelling, hence internal surfaces are not available for exchange reactions. Cation exchange capacity (CEC) is primarily due to pH dependent charge sites on external surfaces. These minerals have low CEC values.

b. Illites and vermiculites. If the dominant interlayer cation is K^+ or NH_4^+ the minerals will not swell and they will be very similar to the micas. That is the interlayer cations will be "fixed" and CEC will be low. If the dominant interlayer cation is Ca^{2+} or some other cation with a nonfavorable fit in the interlayer holes, the mineral will swell and the CEC will be high. In nature, illites will have K^+ as the dominant interlayer cation and occur as nonswelling minerals with low CEC values, while vermiculite will have Ca^{2+} or Mg^{2+} as the dominant interlayer cation. Hence vermiculite occurs as a swelling type mineral with a high CEC.

c. Smectites. Interlayer cations are not fixed regardless of their hydrated size. This means that all of the permanent charge contributes to the CEC, unlike the micas where the permanent charge is satisfied by "fixed" cations. The smectites are swelling minerals, hence the internal as well as the external surfaces are available for exchange reactions. The swelling nature and the charge results in medium to high CEC values. In general at maximum CEC about 80% of the CEC is due to isomorphous substitution and 20% due to pH dependent charge.

8.5 PERCENT BASE SATURATION.

Percent base saturation is that ratio of basic cations to acidic cations on the exchange complex. Acidic cations are cations such as the H^+ ion and Al^{3+} that produce acidic solutions when added to water. Basic cations are cations such as K^+, Na^+, or Ca^{2+} that produce neutral or basic solutions when added to water. In soils work the only acidic cations normally encountered are the hydrogen and aluminum cations.

$$\% \text{ base saturation} = \text{cmol}_c \text{ of basic cations/cmol}_c \text{ of cation exchange capacity.}$$

Percent base saturation is closely related to the pH of the soil solution. This relationship will be examined in detail in the pH laboratory.

8.6 DETERMINATION OF CATION EXCHANGE CAPACITY.

Basic steps:

1. Replace cations originally on exchange complex by washing with potassium acetate.

2. Wash the excess potassium ions out of the soil with alcohol so that the only potassium ions remaining in the soil are those absorbed on the exchange complex.

3. Elute the exchangeable potassium from the soil with ammonium acetate.

4. Determine concentration of the eluted potassium using the emission mode of the atomic absorption spectrometer.

Laboratory procedure.

1. Determination of CEC.

a. Weigh out and record exact weight of approximately 5 grams of soil. Instructor will furnish air-dry moisture contents for soils.

b. Transfer soil into 100 centrifuge tube.

c. Add 20 ml of 1N potassium acetate ($KC_2H_3O_2$).

d. Stopper and shake for 1 minute, unstopper and rinse soil on stopper and tube sides into tube with a squirt bottle (Notes: use a limited amount of water).

e. Centrifuge for 5 minutes or until clear.

f. Discard clear liquid.

g. Repeat steps c through f one additional time. **Note**, these steps saturated the soil with potassium ions.

h. Add 20 ml methyl alcohol (CH_3OH).

i. Stopper and shake until soil pellet is resuspended, remove and rinse stopper as in step d.

j. Centrifuge for 5 minutes or until clear.

k. Discard clear liquid.

l. Repeat steps h through k 2 times. **Note**, these steps wash the excess potassium ions out of the soil so that only exchangeable potassium remains.

m. Add exactly 20 ml of 1N ammonium acetate.

n. Stopper and shake until the soil pellet is resuspended, remove and rinse stopper.

o. Centrifuge, pour supernatant (clear liquid) into a clean, dry 100 ml beaker.
Save the supernatant!

p. Repeat steps m through o one time, combine supernatants into the same 100 ml beaker.

q. Dilute the supernatant according to the directions provided by your laboratory instructor.

r. Determine the concentration of potassium (µg/ml) in the supernatant according to your lab instructor's directions using the atomic absorption spectrometer or flame photometer.

Data - CEC experiment

Air-dry weight of soil sample	_______
Volume of extraction solution	_______
Dilution factor	_______
Absorbance (atomic absorption spectrometer)	_______
µg K^+/mL (from standard curve)	_______

Calculations - CEC experiment.

a. µg K^+/g soil = (µg K^+/ml x dilution factor x volume of extracting solution) / weight of soil.

dilution factor = provided by your laboratory instructor

b). CEC = $cmol_c$/kg soil; 1 cmol K^+ = 1 $cmol_c$; 391,000 µg K^+ = 1 cmol K^+

CEC = µg K^+/g soil x 1000 g/kg x 1 $cmol_c$/391,000 µg K^+

2. Flocculation experiments.

a. Pipette 5 ml of the dispersed clay provided by the T.A. into 5 test tubes.

b. Add 1 ml of the following salt solutions to different 5 ml samples in the test tubes.

1N KCl, 1N $CaCl_2$ and distilled water.

c. Label each tube to identify the salt solutions added.

d. Swirl the tubes to mix the salt solutions with the suspension.

e. Observe and record the rate of flocculation and the size of the floccules.

Rank the cations as to their effectiveness in flocculating the clay suspensions. Discuss the reasons for the observed differences.

Data flocculation experiment.

1N KCl Observations ____

1N $CaCl_2$ Observations ____

Distilled water Observations ____

3. Physical properties of clays:

   a. Add two grams of kaolinite to a plastic beaker.

   b. Fill a 100 ml graduated cylinder with tap water.

   c. Add water slowly from the cylinder to the clay in the beaker until you get a smooth creamy mixture.

   d. Record the volume of water used.

   e. Repeat the above procedure with two grams of montmorillonite.

   Discuss the reasons for the different amounts of water needed to produce the smooth creamy mixtures (pastes).

   Data physical properties of clays

   Weight of kaolinite ________ mLs water to make paste _______

   Weight of montmorillonite _______ mLs water to make paste _______

4. Write up the experiments, including a brief introduction, the data, calculations and interpretations of the results.

5. Answer the following questions and include them with your laboratory write-up.

   a. K^+ is used to replace the original exchangeable ions in this experiment. Is it possible to use other cations?

   b. What factors determine the amount and type of cations on the exchange complex.

c. Why are exchangeable cations not held tightly to the colloidal surface?

d. What effect does the presence of CEC in a soil have on the relative loss of cations and anions by leaching?

e. Discuss the difference between an "exchangeable" and a "fixed" cation.

f. Why do only certain types of clays "fix" K^+ and/or NH_4^+?

g. How would you modify this experiment to analyze for exchangeable cations?

h. What is the purpose of using alcohol instead of water to remove the excess K^+ ions out of the soil?

i. Why does Na^+ disperse soils?

j. Explain the effect of ion valence on the flocculation of clays in the flocculation experiments.

LABORATORY 9

SOIL pH AND LIME RECOMMENDATIONS.

Soil pH is one of the most interesting and informative soil properties. Soil pH is a measure of the hydrogen ion concentration in the soil solution. Theoretically, soil pH is the negative logarithm of the hydrogen ion concentration in the soil solution.

$$pH = -\log (H^+)$$

In practice uncertainties in measuring pH associated with dilution of the soil solution and colloidal effects result in pH being an approximate value.

Soil pH is an indicator of soil weathering. Soil pH values reflect the mineral content of the parent material, the length of time and severity of weathering and especially the leaching of basic materials from the soil profile. Factors such as the type of vegetation, annual rainfall, and drainage as well as the activities of man also influence soil pH.

The availabilities of iron, copper, phosphorus, zinc and other nutrients, as well as the toxicities substances are controlled in a large part by soil pH. Some potentially toxic substances in soils, such as aluminum (Al^{3+}) and lead (Pb^{2+}), have little affect on plant growth in alkaline calcareous soils, but are a serious concern at the same soil concentrations in acid soils. Many nutrients, such as phosphorus show their greatest availability in soils with slightly acid to neutral pH values, with marked decreases in availability with increases or decreases in soil pH.

Soil pH is also an indicator of serious soil problems. Soil pH values above 8.5 are indicative of sodic soils. While pH values below 4 suggest the oxidation of reduced sulfur compounds.

9.1 ROLE OF WATER.

Water is one of the most important species in aqueous systems such as soils, not only because it is an excellent solvent but also because of it's role in acid-base reactions. Water will autohydrolyze into a hydronium (H_3O^+) or as more commonly written the hydrogen (H^+) ion and a hydroxyl (OH^-) ion.

$$2H_2O = H_3O^+ + OH^- \qquad Kw = 10^{-14} \text{ at } 25°C$$

or

$$H_2O = H^+ + OH^-$$

Where Kw is the equilibrium constant for the autohydrolysis reaction and is given by:

$$Kw = (H^+)(OH^-) = 10^{-14}$$

The equilibrium constant expression for water has special significance to aqueous systems. This expression and it's logarithmic form, not only establishes the pH scale, and hence the definition of acidic, basic and neutral solutions, but also illustrates the interdependence of (H^+) and (OH^-) concentrations.

In logarithmic form:

$$\log (H^+) + \log (OH^-) = -14$$

In terms of negative logarithms, p = -log and pH = -log (H^+):

$$pH + pOH = 14$$

The equations illustrate that in aqueous solutions the concentrations of the H^+ and OH^- ions can not be varied independently. When one species is increased there must be a corresponding decrease in the concentration of the other, such that the product of their concentration is a constant (Kw). Second, for the pH scale as defined in equation 6, it must be remembered that pH is a negative logarithmic scale. Hence, a change in pH from 5 to 4 is a ten fold increase in H^+ ion concentration and a pH change form 4 to 8 is a ten thousand fold decrease in hydrogen ion concentration.

9.2 MECHANISMS THAT CONTROL SOIL pH.

Table 9-1 Mechanisms that control soil pH.

Soil pH range	Major mechanism(s) operating
2 to 4	Oxidation of pyrite and other reduced sulfur minerals. Dissolution of soil minerals.
4 to 5.5	Exchangeable Al^{3+} and its associated hydroxy ions. Exchangeable H^+.
5.5 to 6.8	Exchangeable H^+. Weak acid groups associated with soil minerals and humic substances. Dissolved CO_{2gas} and aqueous species.
6.8 to 7.2	Weak acid groups on soil organic matter and humic materials.
7.2 to 8.5	Dissolution of solid divalent carbonates ($CaCO_3$s).
8.5 to 10.5	Exchangeable Na^+ under low salt conditions. Dissolution Na_2CO_3s.

Note: For the pH range of 4 to 6.8, soil pH is controlled by the percent of the cation exchange complex occupied by acidic cations (H^+ and Al^{3+})

Oxidation of reduced sulfur compounds.

When soils or deposits that contain reduced forms of sulfur such as iron sulfide are exposed to aerobic environments, conditions are created that favor the oxidation of the reduced sulfur compounds. Associated with the oxidation of the reduced sulfur minerals is the production of H^+ ions and their release to the soil solution.

$$2FeS_{pyrite} + 3.5O_{2gas} + H_2O = \alpha\text{-}Fe_2O_{3hematite} + 4H^+ + 2SO_4^{2-}$$

The reduced sulfur compounds are oxidized by chemotropic bacteria such as thiobacillus. The chemical energy released by these oxidations is used by the bacteria to drive their metabolic processes. The initial process of sulfur oxidation in neutral soils is carried out by bacteria , but once the soil has been acidified to pH values around 3.5 or less the oxidation of the sulfide ions can be coupled with the reduction of ferric iron and the process becomes chemical (abiotic) in nature.

$$Fe^{3+} + S^{2-} + 4H_2O = Fe^{2+} + SO_4^{2-} + 8H^+$$

The source of the acidity in these soils is the oxidation of reduced sulfur. The total amount of acidity that can potentially be produced is determined by the quantity of reduced sulfur minerals in the soil. The rate at which the acidity is produced is governed by the rates of the biological and chemical mechanisms.

If the pH values of these soils are determined and compared to the amount of acid being produced, it is apparent that some buffering mechanism in the soil is operating and reacting with the H^+ and preventing the establishment of extremely low pH values. In acid sulfate soils, it has been shown that more that 98% of the acid released during pyrite oxidation reacts with soil minerals and is neutralized, 1 to 2% reacts with dissolved alkalinity (HCO_3^-) and less than 1% remains in the soil solution as free acid.

$$Al(OH)_{3gibbsite} + H^+ = Al(OH)^{2+} + H_2O$$

The equation showing gibbsite reacting with a H^+ ion is an example of how a soil mineral can neutralize the H^+ ions produced by sulfur oxidation. Soil minerals in addition to gibbsite can also be dissolved in response to the elevated H^+ ion levels and contribute to overall soil buffering.

Exchangeable aluminum.

The trivalent aluminum ion (Al^{3+}) is the cation of a weak base and as such has the potential to hydrolyze water producing hydrogen ions. The combination of the aluminum ion, its hydrolysis species and the insoluble weak base $Al(OH)_{3amorphous}$ represents a vigorous buffer system for soils in this pH range. Consider the following species of aluminum and their chemical equilibria. Additional species of aluminum exist, but these species will be sufficient to illustrate how pH is controlled in this pH range.

1. Soluble species.

Al^{3+}, $AlOH^{2+}$, $Al(OH)_2^+$

2. Solid species.

$Al(OH)_{3amorphous}$

$$Al(OH)_{3amorphous} + H^+ = Al(OH)_2^+ + H_2O \qquad K_1 = 10^{-0.081}$$

$$Al(OH)_2^+ + H^+ = AlOH^{2+} + H_2O \qquad K_2 = 10^{4.7}$$

$$AlOH^{2+} + H^+ = Al^{3+} + H_2O \qquad K_3 = 10^{5.0}$$

In neutral soils the solubility of aluminum solid species is very low. For example the concentration of $Al(OH)_2^+$ in equilibrium with amorphous $Al(OH)_3s$ at a pH of 7 is equal to $8.3x10^{-7}$ M. By the time the soil has acidified to pH 5 the concentration of $Al(OH)_2^+$ has increased to $8.3x10^{-5}$ M. This represents more than a hundred fold increase in solubility. The dissolution of the solid $Al(OH)_{3amorphous}$ as well as the shift from one aqueous species of aluminum to another consumes H^+ ions acting as a buffer to slow the acidification process.

Once the soil has acidified, Al^{3+} and the other soluble species not only will have increased in the soil solution, but will also have accumulated on the soil colloids as exchangeable cations. When the soil has reached this state, the aluminum system not only buffers the soil against decreases in soil pH, but also against increases in pH. If a basic material is added to the soil the species neutralize the OH^- ions as shown by the reverse of equations 10-12.

$$Al^{3+} + OH^- = AlOH^{2+} \qquad K = 1/KwK_3$$

or

$$AlOH^{2+} + OH^- = Al(OH)_2^+ \qquad K = 1/KwK_2$$

The pH of the soil is buffered not only by the aluminum species in the soil solution, but also by the aluminum species on the cation exchange complex. The aluminum species associated with both phases must be converted back to insoluble solid species to increase pH up to the level (pH > 6.2) usually required for crop production. Aluminum species on the cation exchange complex along with exchangeable H^+ constitute the exchangeable acidity. The control of soil pH by exchangeable acidity will be discussed in the next section.

Exchangeable hydrogen (acidity).

The concentration of a particular type of cation on the cation exchange complex relative to the concentration of all other types of cations on the exchange complex governs the concentration of that type of cation in the soil solution. There is an equilibrium established between the cations on the exchange complex and the cations in the soil solution.

[exchangeable cation] = [cation in soil solution] K_{eq}

[exchangeable H^+] = [H^+ in the soil solution] K_{eq}

Where K_{eq} is the equilibrium or exchange constant for the reaction. The percentage of the exchange complex that is occupied by a particular cation determines the concentration of that cation in the soil solution. Figure 9-1 illustrates the relationship between hydrogen ions on the exchange complex (exchangeable acidity) and the concentration of hydrogen ions in the soil solution (active acidity).

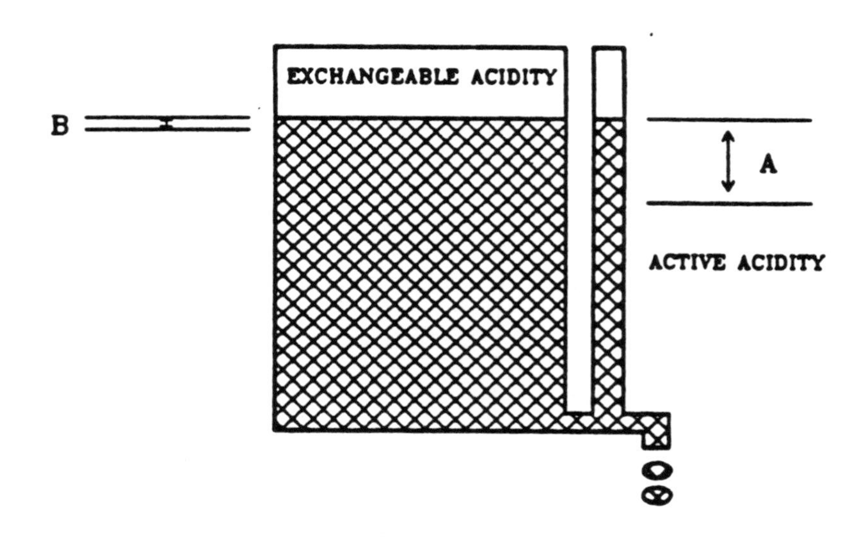

Figure 9-1 Diagrammatic relationship between exchangeable and active acidity.

When liquid is withdrawn from the standpipe, the level in the standpipe decreases and stays low until the value is shut. Once the value is shut the level in the standpipe will increase back to the level in the large tank. Since the volume of the tank is much larger than the standpipe, it is necessary to remove many standpipe volumes before the change is reflected in the level of liquid in the tank. The volume of the large tank is a model of the amount of H^+ ions on the exchange complex, while the volume of the standpipe is a model of the amount of H^+ ions in the soil solution. The level of H^+ ions in the soil solution is controlled by the level of H^+ ions on the exchange complex. When H^+ ions are added or removed from the soil solution, the exchangeable acidity is affected very little and the original level of H^+ ions in the soil solution is restored. The exchangeable H^+ ions buffer the level of H^+ ions in the soil solution.

Percent base saturation is the ratio of basic (nonacidic) cations on the exchange complex to the total cation exchange capacity of the soil.

% base saturation = [conc. basic cations ($cmol_c/kg$)]/CEC($cmol_c/kg$)

When percent base saturation is high the concentration of basic cations (i.e Ca^{2+}, K^+, Mg^{2+} etc) on the exchange sites is high and the concentration of acidic cations (H^+, Al^{3+} and the related hydrolysis species) is low. Since the level of ions on the exchange complex governs the level of ions in the soil solution. When percent base saturation is high the concentration of basic cations in the soil solution is high and the concentration of acidic cations is low. When base saturation is low the concentration of basic cations in the soil solution is low and the concentration of acidic cations in the soil solution is high. Figure 9-2 illustrates the relation between soil pH and percentage base saturation.

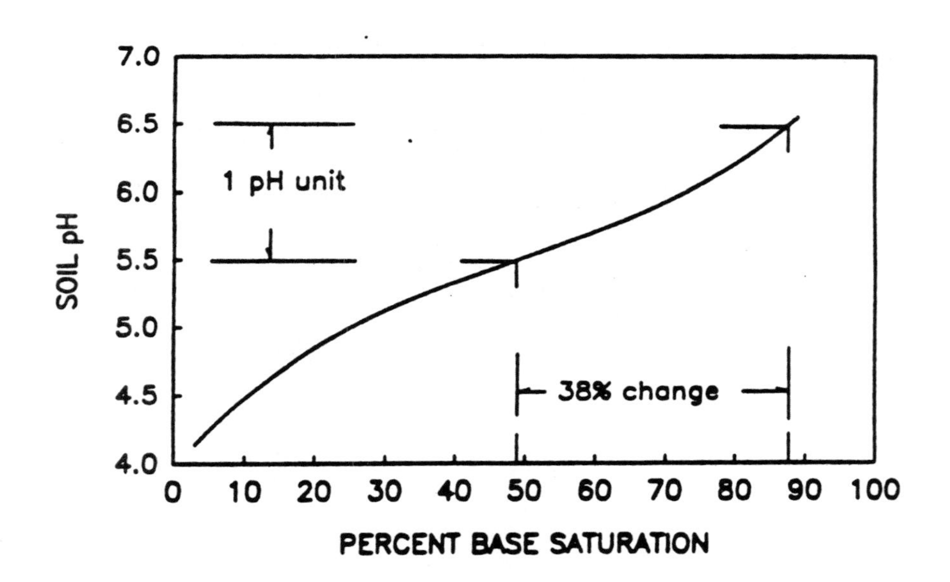

Figure 9-2 Titration curve for an ideal soil.

Figure 9-2 shows that when the soil solution has a pH of 5.5, the corresponding percent base saturation is equal to 48%. When the pH of the soil is 6.5 the corresponding percent base saturation is 86%. In order to change the pH of the soil from 5.5 to 6.5 as would be the case in a liming operation, not only must the hydrogen ions in the soil solution be neutralized, the percent base saturation must be changed form 48% to 86%. That is the amount of hydrogen equal to 38% (86% - 48%) of the CEC must be removed from the exchange complex and neutralized and replaced with basic cations. To permanently change the H^+ ion concentration in the soil solution, the ratio of H^+ ions to other cations on the exchange complex must be changed.

Weak acid groups on soil clays, hydrous oxides and soil humic substances.

The dissociation of weak acid groups such as -AlOH and -SiOH groups at the edges of soil clays, -FeOH and -AlOH groups associated with hydrous oxides and phenolic and carboxyl groups of soil organic matter are important sources of H^+ ions and contribute to the overall buffering of soils. In addition these groups are responsible for the pH dependent charge of soils. These functional groups are tightly bound to and very specific for the H^+ ion in acid soils.

$$-AlOH = -AlO^- + H^+$$

$$-SiOH = -SiO^- + H^+$$

$$-FeOH = -FeO^- + H^+$$

$$R\text{-}OH = R\text{-}O^- + H^+$$

$$R\text{-}COOH = R\text{-}COO^- + H^+$$

Where R is the organic molecule to which the function group is attached. In acid soils the reactions are shifted to the left towards the undissociated groups because of the concentration of H^+ ions in the soil solution. As the soil pH is raised the groups dissociate becoming an important source of H^+ ions in near neutral soils. If H^+ ions are added to the soil the weak acid groups buffer the soil by reacting with the H^+ ions and if OH^- are added the additional groups can dissociate producing H^+ ions to react with the added OH^- ions.

The -AlOH, -SiOH and -FeOH tend to dissociate over a fairly narrow pH range, while the groups associated with soil organic matter begin to dissociate at lower pH values and continue over a wider pH range.

Carbonic acid and aqueous carbon dioxide.

Carbon dioxide gas and the associated aqueous species comprise a very important source of H^+ ions and an important buffer system in soils and natural waters. The atmosphere above the soil contains a partial pressure of CO_2gas equal to 0.0003 atm.. This gas dissolves in precipitation to eventually produce carbonic acid (H_2CO_3). Although carbonic acid is a weak acid, it is a constant source of H^+ ions. The following equations detail the aqueous chemistry of CO_2gas.

$$CO_2gas = CO_2aq \qquad K_1 = 10^{-1.41}$$

$$CO_2aq + H_2O = H_2CO_3 \qquad K_2 = 10^{-2.62}$$

$$H_2CO_3 = HCO_3^- + H^+ \qquad Ka_1 = 10^{-3.76}$$

$$HCO_3^- = CO_3^{2-} + H^+ \qquad Ka_2 = 10^{-10.25}$$

Calcareous soils.

The pH of calcareous soils is due to the presence of salts of a weak acid and a strong base. The salts or solid phases may be calcite or aragonite, different crystal arrangements of $CaCO_3s$, Mg contaminated calcium carbonates, magnesite $MgCO_3s$ or dolomite $CaMg(CO_3)_2s$. Regardless of the solid phase present, when the salt dissolves the anion (CO_3^{2-}) of a weak acid carbonic acid is released into the soil solution. For a soil containing calcite:

$$CaCO_{3}\text{calcite} = Ca^{2+} + CO_3^{2-} \qquad Ksp = 10^{-8.3}$$

The carbonate ion will hydrolyze water as shown by the following equation.

$$CO_3^{2-} + H_2O = HCO_3^- + OH^- \qquad K = Kw/Ka_2$$

Sodic soils.

The high pH of sodic soils is due to a complex combination of exchange reactions and the autohydrolysis of water. The sodium ion (Na^+) is a monovalent ion that can be easily replaced on the exchange complex by H^+ ions. This exchange of H^+ for Na^+ can occur readily under certain condition and result in increased pH values in the soil.

$$\text{clay-}Na^+ = H^+ + OH^-$$

$$\text{clay-}H^+ = Na^+ + OH^-$$

Consider two situations, first, a soil solution containing a high concentration of dissolved salts and second, a soil solution with low level of dissolved salts such that the H^+ and OH^- ions resulting from the autohydrolysis of water are significant ions. In the first case, when exchange of Na^+ occurs it most probably would be for a basic cation such as K^+, Ca^{2+} or Mg^{2+} since they would be the major cationic species present in the soil solution. This would have no effect on the pH of the soil solution since there would be no change in the balance of H^+ and OH^- ions in the soil solution. In the second case, H^+ because of the low salt environment of the soil solution is a major cationic species. When exchange for the Na^+ ion occurs the exchange will often be with the H^+ ion. The exchange of H^+ for Na^+ results in a decrease in the H^+ ion concentration relative to the OH^- ion concentration and hence, an increase in the pH of the soil solution. The Na^+ ion is not readily exchanged for by the H^+ ion until there is a fairly large portion of the exchange capacity occupied by the Na^+ ion. Research has shown that the exchange of H^+ for Na^+ does not become a problem until more than 15% of the exchange complex is occupied by Na^+ and unless the dissolved salt content of the soil solution is low enough that the conductivity of the soil's saturation extract is less than 4 dS/m.

If the pH of a sodic soil becomes high enough a second mechanism becomes involved and buffers the soil. As the pH of the soil increases CO_2gas in the atmosphere dissolves in the soil solution. Because of the high pH of the soil the aqueous-CO_2 system is shifted towards the carbonate species (CO_3^{2-}), resulting in a high concentration of the carbonate ion. When the CO_3^{2-} ion concentration multiplied by the concentration of the Na^+ ion squared exceeds the Ksp of sodium carbonate (Na_2CO_3s), sodium carbonate will precipitate and buffer the pH of the soil solution.

$$Na^+ + CO_3^{2-} = Na_2CO_{3}s$$

$$Ksp = (Na^+)^2(CO_3^{2-})$$

Once the solid phase has formed it will buffer the soil at a pH value close to 10.5 by a mechanism very similar to that controlling pH in calcareous soils.

9.3 MEASUREMENT OF SOIL pH.

Soil pH as used in this text refers to the negative logarithm of the hydrogen ion concentration (H^+) in the soil solution. Predictions about the chemistry of a soil constituent are often based on the pH of the soil solution. In actual practice these predictions are based on measured pH values and measured pH values may or may not be reliable estimates of the actual pH of the soil solution. McLean (1982) identified the following factors that may influence the accuracy of the measured pH:

1. The nature and type of inorganic and organic constituents that contribute to soil acidity.

2. The soil to solution ratio used to measure the pH.

3. The salt content of the diluting solution used to achieve the desired soil to solution ratio.

4. The CO_2gas content of the soil and solution.

5. Errors associated with the standardization of the equipment used to measure pH.

The measurement of soil pH generally requires that additional water or salt solution be added to the soil. The most common soil to solution ratio used to measure pH with a pH electrode is a 1:1 dilution, a 2:1 dilution is often used in soil testing, as are 1:10 dilutions.

The addition of water or salt solution can result in the (H^+) in the final diluted solution being different from the (H^+) concentration in the original soil solution. Upon dilution, weak acid groups associated with soil organic matter as well as soil minerals may associate or dissociate. The salt content of the diluting solution can result in the release of H^+ ions from the exchange complex. Depending upon the nature and type of colloids and weak acid groups present, the effect of dilution may differ from one soil to another, making consistent correction difficult if not impossible. The use of 0.01 M $CaCl_2$ has been recommended to minimize the effects of dilution, since this concentration of salt is generally considered to represent the salinity of a "normal" soil solution in the field.

Dilution of the soil solution may also result in differences in the type and amount of dissolved gases. Gases such as CO_{2gas} are present in the soil in higher concentrations than in the bulk atmosphere. Hence dilution and stirring as used in the actual measurement of soil pH can result in the loss of such gases and differences in the measured and actual pH of the soil.

Soil pH measurement is normally made using either electrometric or colorimetric techniques. Colorimetric techniques are based on structural changes of chromophore groups of organic compounds in response to the pH of the soil solution. Normally the pH indicator is added to the soil and the resulting color compared with a color chart. Problems are encountered related to the native color of the soil interfering with the determination of the color of the pH indicator.

Electrometric methods, based on the pH sensitive glass electrode are generally accepted as the standard method for determining soil pH. The glass electrode consists of a thin glass membrane containing a solution of HCl and KCl in contact with an internal AgCl or $Hg\text{-}Hg_2Cl_2$ electrode which is connected to the pH meter. The thin glass membrane is in contact with external soil solution (diluted soil solution) of unknown (H^+) ion concentration and in contact with the internal HCl-KCl solution of known (H^+) ion concentration.

The glass membrane has cation exchange properties, with a particularly high degree of sensitivity to H^+ ions. When the glass membrane is placed in contact with an external solution a potential is developed across the membrane and is sensed by the internal AgCl or $Hg\text{-}Hg_2Cl_2$ electrode.

The measurement of pH using the glass electrode requires that there be a complete conducting path for the electrons. The circuit is completed using a second external reference electrode, usually a $Hg\text{-}Hg_2Cl_2$ (calomel) which is place in the unknown external solution and connected to the pH meter.

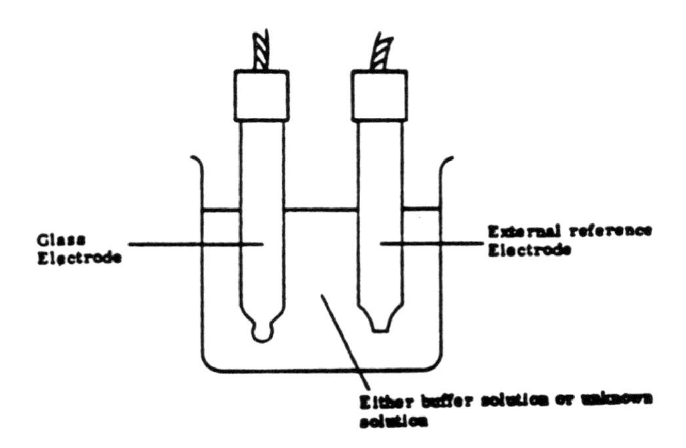

Figure 9-3 Typical arrangement of electrodes for a pH meter.

9.4 DETERMINATION OF LIME REQUIREMENTS.

The amount of lime that must be added to a soil to change its pH is a function of the soils buffering capacity. Two soils with the same distilled water pH values may or may not need the same amount of limestone to adjust their pH values.

In the pH ranges found for most temperate region agricultural soils buffering is controlled by the relationship between percent base saturation and soil pH. Hence the amount of limestone or other liming material needed to adjust the pH of the soil is a function not only of the soils pH but also its cation exchange capacity.

In order to adjust the pH of a soil the hydrogen ions in the soil solution and on the exchange complex must be neutralized. If two soils require the same change in pH, for example from 5.5 to 6.2 the soil with the greatest CEC will require the most liming material.

The two methods most commonly used to determine lime requirements in the midwest are the SMP buffer method and the Illinois method.

SMP buffer method.

This method is based on the change in the pH of a buffer solution following the addition of soil. The buffer is prepared from p-nitrophenol, triethanolamine, calcium acetate, calcium chloride, potassium dichromate and water and has an initial pH of 7.5. Upon addition of the buffer to the soil the hydrogen ions in the soil both in the soil solution and on the exchange complex neutralize the buffer and result in a decrease in its pH from 7.5. The amount of decrease in pH of the buffer is a function of the total amount of hydrogen ions in the soil, that is its buffering capacity. The following table gives the amount of typical limestone* needed to bring a mineral or organic soil to desired pH value.

Table 9-2 Amounts of lime (tons/acre 8" of soil) needed to bring the soil to desired pH.

	Desired soil pH ------- Mineral soils		------------	Organic soils
Soil buffer reading	7.0	6.5	6.0	5.2
6.8	1.4	1.2	1.0	0.7
6.6	3.4	2.9	2.4	1.8
6.4	5.5	4.7	3.8	2.9
6.2	7.5	6.4	5.2	4.0
6.0	9.6	8.1	6.6	5.1
5.8	11.7	9.8	8.0	6.2
5.6	13.7	11.6	9.4	7.3
5.4	15.8	13.4	10.9	8.4
5.2	17.9	15.1	12.3	9.4
5.0	20.0	16.9	13.7	10.5
4.8	22.1	18.6	15.1	11.6

* Ag limestone of 90% plus CCE and a fineness of 40% < 100 mesh, 50% < 60 mesh, 70% < 20 mesh and 95% < 8 mesh.

Illinois method.

The Illinois method is based on incubations of various textured soils with a typical limestone. Soil test form D-2 gives two charts that allow the determination of soil pH from distilled water pH values and estimation of the soils CEC from color and texture measurements.

Chart I is for cash grain cropping systems where no legume other than soybeans is to be grown. Cash grain cropping systems require a minimum pH of 6.0. Chart II is for cropping systems with alfalfa, clover or lespedeza. These cropping systems require a minimum pH of 6.5.

Soil Test Form D-2. LIME -- Map and Interpretation[1]

Charts I and II (limestone needed by grain farming systems and cropping systems with alfalfa, clover or lespedeza) are based on these assumptions:

1. A 9 inch depth of plowing. for each inch less, the limestone requirement may be reduced by 10 percent.

2. Typical fineness limestone - 10% > 8 mesh

   30% 8 - 30 mesh
   30% 30 - 60 mesh
   30% < 60 mesh

3. A calcium carbonate equivalent of 90%.

If these assumption due not apply to your situation, adjust the limestone rate accordingly.

STEPS TO FOLLOW

1. Use Chart I for grain systems and Chart II for alfalfa, clover or lespedeza.

2. Decide which class (estimate of CEC) fits your soil.

   A. Silty clays and silty clay loams (dark).
   B. Silty clays and silty clay loams (light and medium).
      Silt and clay loams (dark).
   C. Silt and clay loams (light and medium), sandy loams (dark), loams (dark and medium).
   D. Loams (light), sandy loams (light and medium), sands.
   E. Mucks and peat (organic soils).

The amounts of soil organic matter that corresponds to color terms are: light < 2.5%, medium 2.5 - 4.5%, dark > 4.5%.

3. Find your soil's pH along the bottom of the chart that fits your cropping system.

4. Follow up the vertical line until it intersects the diagonal line A, B, C, D, or E that fits your soils class (estimated CEC).

5. Read the suggested rate of application along the right side of the chart that you are using.

---

[1] Department of Agronomy, University of Illinois, Urbana Illinois.

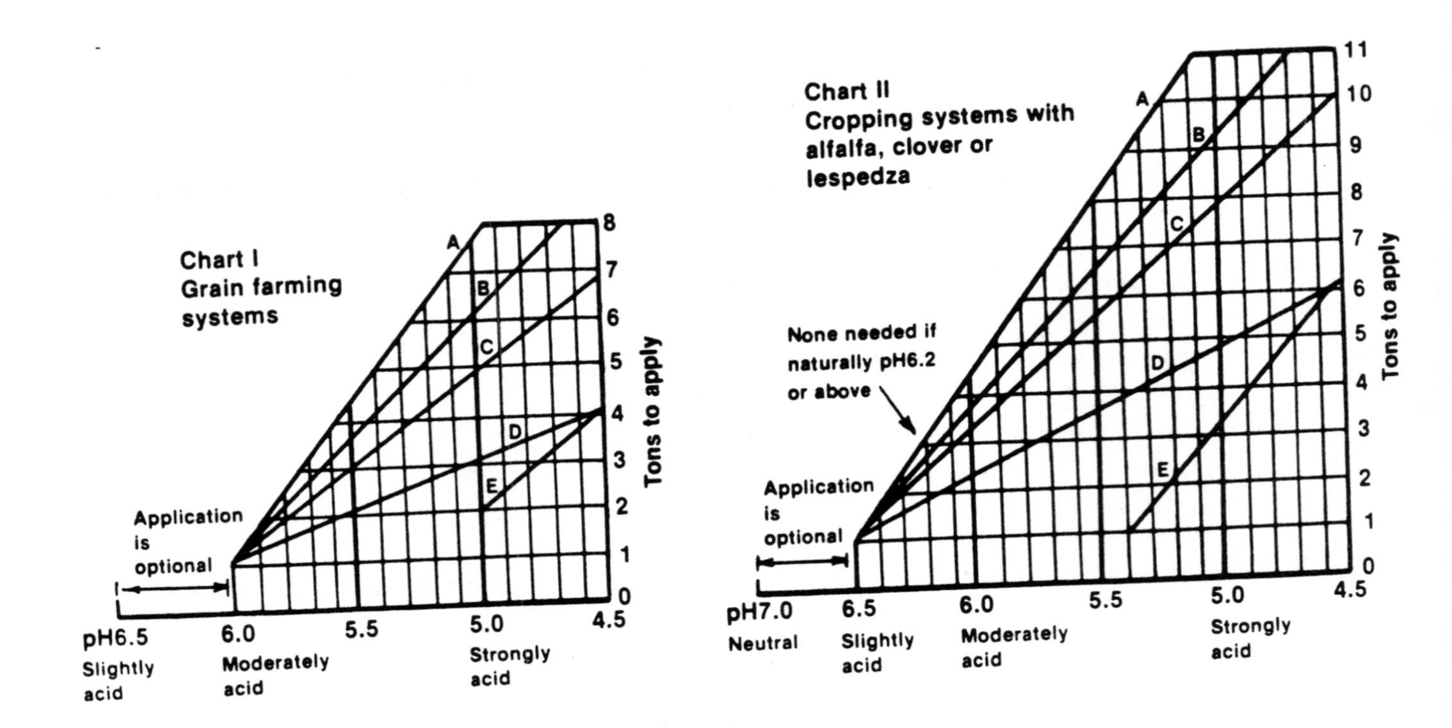

Information to help you plan a liming program

Reasons for liming. Liming acid soils improves the growth and yields of most crops because:

1. It reduces the solubility of manganese and aluminum that are present in strongly acid soils in amounts large enough to be toxic especially, to alfalfa and clover crops.
2. It improves the soil for microorganisms that facilitate decay of plant residues, thus releasing more nitrogen and phosphorus for the crop.
3. It favors the growth of beneficial bacteria such as the symbiotic nitrogen fixing bacteria that form nodules on legumes (i.e., soybeans, clover and alfalfa).
4. The best balance in availability of micronutrients is found in soils that are only slightly acid (~6 to 6.5).
5. Phosphorus in more available in soils that are near neutral in acidity that it is in more acid soils.

Suggested pH goals. For cropping systems with alfalfa, clover or lespedeza in the rotation maintain a pH of 6.5 or above. But if the soils have a pH of 6.2 or above without ever having been limed, with neutral soil just below the plowing depth it probably will not be necessary to apply limestone.

For cash grain cropping systems (no legumes except soybean) maintain a pH of at least 6.0. Farmers may choose to raise the pH to still higher levels. After initial investment, it cost little more to maintain a pH of 6.5 than one of 6.0.

Liming materials. Nearly all the liming material used in Illinois is agricultural ground limestone, commonly called agstone or lime. The active ingredient in the limestone may be calcite (calcium carbonate), magnetite (magnesium carbonate) or dolomite (calcium-magnesium carbonate). Calcitic limestone is predominately calcium carbonate, while dolomitic limestone contains either magnetite or dolomite. The maximum magnesium carbonate content of Illinois limestone is about 45%. There is no area in Illinois where one type of limestone is preferred over the other.

The two main characteristics that determine the value of limestone are its calcium carbonate equivalent (CCE) and its fineness. The calcium carbonate equivalent (neutralizing power) of limestone is when everything of liming value in the limestone is expressed in terms of the amount of pure calcium carbonate that will neutralize the same amount of acidity. The CCE of limestone sold in Illinois ranges from about 65 to 109. The higher the CCE the more valuable the limestone. Limestone that has a lower CCE may be just as good a bargain if it is priced in line with the difference in CCE.

Size fraction	Years after application		
	1	4	8
Through 60 mesh	100	100	100
30 to 60 mesh	50	100	100
8 to 30 mesh	20	45	75
Over 8 mesh	5	15	25

The **fineness**, or size, of limestone particles largely determines the rate at which the limestone particles dissolve and hence neutralize soil acidity. The following table shows the percent of added limestone particles of different sizes that dissolve 1, 4 and 8 after application.

If you are liming a strongly acid soil shortly before seeding alfalfa or clover, the values for 1 year in the table are the best guide. If lime is applied to sod before plowing for corn, the 4 year values are the best to use. If the fields have been limed in the past and a long-term maintenance application is to be made, the 8 year figures are satisfactory.

When to lime. Fortunately limestone may be applied at many points in the crop rotation. It is best to lime soils for the first time 6 months to a year ahead of alfalfa or clover seeding, but in emergencies limestone that contains a high proportion of 60 mesh material can be disked in (not plowed under) just ahead of seeding. If the amount of limestone is 6 tons or more and the initial cost is a factor, apply two-thirds the first time and the rest two to four years later.

Retesting limed fields. Soil pH tests within 2 years after liming are unreliable. When lime has been applied according to soil test, alfalfa, clover and other crops will grow well even though the soil still test more acid than is desired. There is little reason to retest a well-limed field more often than every 4 years. Eight to10 years is often enough to retest fields with a naturally high soil pH. Where high rates of nitrogen (150 lbs/acre/yr) are applied, retesting is suggested every 4 years because nitrogen fertilizer increases soil acidity.

Laboratory procedures.

1. Measurement of soil pH with indicators.

   a. Rinse spot plate with deionized water.

   b. Fill 1/3 depression with pulverized soil.

   c. Add indicator dye (usually start with chlorophenol red or bromothymol blue). An enough dye should be added to saturate the soil and have some excess solution on top of the soil.

   d. Gently swirl the spot plate for 1 minute to allow the dye to interact with the soil.

   e. After 5 minutes tip the plate slightly so that you can see the color of the indicator solution against the white porcelain plate.

   f. Match the solution color to that on the color chart. If the color matches any of the last 2-3 color chips at the high or low end of the chart select the next indicator and repeat the test. If the color matches a center chip record the pH.

Data - soil pH using indicators solutions.

Soil 1 indicator ___________ color ___________

indicator ___________ color ___________

indicator ___________ color ___________

final pH ___________ (the result of the above indicators)

Soil 2 indicator ___________ color ___________

indicator ___________ color ___________

indicator ___________ color ___________

final pH ___________ (the result of the above indicators)

2. Measurement of soil pH with a glass electrode pH meter in distilled water and 0.01 M $CaCl_2$ solution.

   The laboratory instructor will demonstrate the use and calibration of the pH meters in the laboratory.

   a. Place 10 grams of soil into two plastic beakers (10g/beaker).

   b. Add 20 ml distilled to one beaker and 20 ml 0.01 M $CaCl_2$ solution to the other.

   c. Stir intermittently for 20 minutes.

   d. Read the pH values on a standardized pH meter and record as distilled water and salt pH values.

   e. Save the distilled water samples for use with the SMP buffer method of determining lime requirement.

   Data - soil pH using pH meter and lime requirement Illinois method

   Soil 1 water pH ___________ salt pH ___________

   Soil 2 water pH ___________ salt pH ___________

   Determine lime requirement using the water pH and Charts I and II from the Illinois method.

3. SMP buffer method for lime requirement.

   a. Add 20 ml of the SMP buffer solution to the soil samples saved from the distilled water pH measurement.

   b. Stir intermittently for 20 minutes.

   c. Determine the pH on the standardized pH meter and record value.

   d. Rinse the electrodes and store in buffer storage solution.

   Data - lime requirement - SMP buffer method.

   Soil 1 SMP buffer pH _______ Soil 2 SMP buffer pH _______

Determine the lime requirement of the soil(s) from the measured SMP buffer pH and Table 9-2.

4. Answer the following questions:

   a. Most agricultural soils in the midwest have pH values in the range of 6.0 to 6.5. What mechanism(s) are responsible for maintaining soil pH in this range?

   b. Why do calcareous soils have pH values in the range of 8.2 to 8.5?

   c. If a strong acid such as HCl is spilled on an acid soil (pH = 4.5), the pH will immediately be lowered to values around 1 or 2. After a period of time the soil pH values will slowly raise back up to around 4.5. Explain.

   d. What do you measure when you determine a soil's pH value?

   e. Is $CaSO_{4}$gypsum a suitable liming material?

   f. Why do sodic soils have high pH values, but not saline-sodic soils since the both contain high exchangeable sodium.

   g. Is high levels of exchangeable $Al3^{+}$ in acid soils the result or the cause of soil acidity?

   h. Why are Ca^{2+}, Na^{+}, Mg^{2+} and K^{+} etc. called basic cations?

   i. KCl fertilizer is added to a soil. Considering all aspects of weak acid-base chemistry, plant uptake and leaching, how will this affect soil pH?

   j. What is the difference in a calcitic and a dolomitic limestone?

   k. Besides carbonate minerals, what other materials can be used for liming?

   l. How does the SMP buffer pH differ from soil pH?

   m. What are the benefits associated with liming an acid soil?

5. Write-up the experiments, including a brief introduction, the data, calculations and interpretations of results. Include answers to the questions at the end of your write-up.

LABORATORY 10.

BIOLOGICAL ACTIVITY IN SOILS.

Soil contains thousands of species of organisms in a variety of sizes, shapes and functions. Soil organisms are responsible for many important soil processes, ranging from the decomposition of plant and animal residues to the fixation of atmospheric nitrogen (N_2gas) into biologically available forms.

The biological activity experiment is designed to demonstrate:

1. The effect of the C:N ratio of an added organic residue on soil nitrogen levels.

2. The effect of soil pH on chemotrophic oxidation of NH_4^+ to NO_3^- (nitrification).

3. The effect of soil moisture on nitrification

10.1 DECOMPOSITION OF ORGANIC MATERIALS IN SOILS.

One of the most important functions of soil microbes is the decomposition of organic residues produced by plant, animals and other microbes. The decay process releases the nutrients contained in the residues and produces carbon dioxide gas completing carbon cycle (Figure 1). The organisms responsible for the decay process are heterotrophic organisms that obtain both there energy and carbon from the decomposition process.

During the decay process organic materials are partially converted in to humus (soil organic matter), some of the organic materials are assimilated by the microorganisms to make new cell material, but most of the residues are oxidized to produce energy for the organisms. The ability decompose organic residues and the amount of humus, cell material or energy produced is related to the nitrogen content of the residue and the nitrogen available in the soil to support the decomposition process.

Although the carbon content of plant residues varies, it closely approaches 40 percent. In these same residues the nitrogen content varies considerably more. Corn cobs or oat straw have vary low nitrogen contents giving them very wide C:N ratios close to 60:1. While residues from alfalfa contain significantly more nitrogen and hence have much narrow C:N ratios of approximately 20:1.

The significance of a residues C:N ratio is related to the amount of **soil** nitrogen that must be utilized to decompose the residue. Microbial tissue has a C:N ratio of approximately 8:1 or less. For a microbe to decompose a residue such as corn cobs or oat straw with a wide C:N ratio soil nitrogen must be absorbed.

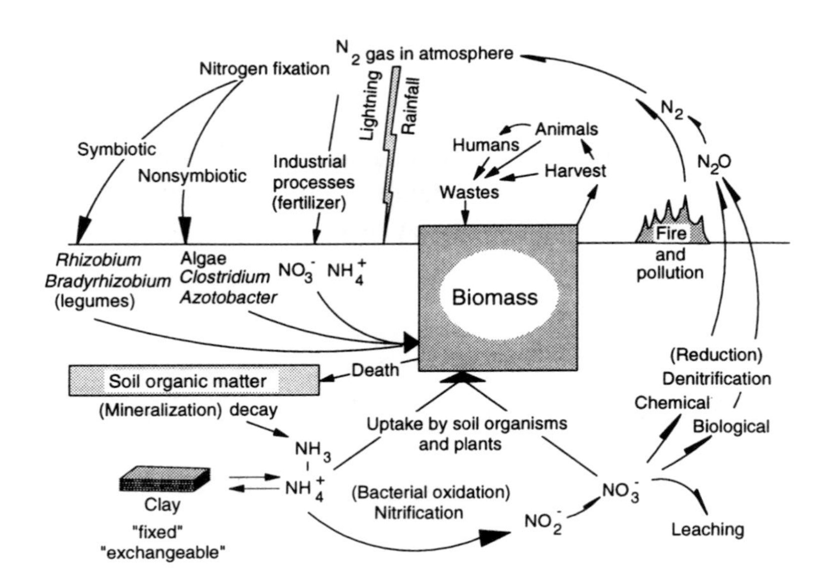

Figure 10-1 The biological nitrogen cycle.

The C:N ratio of humus varies in the range of 10:1 to 15:1, hence amount of humus production is also dependent upon utilization of soil nitrogen. If a residue with a wide C:N ratio such as corn cobs is added to the soil, the availability of soil nitrogen will determine the amount of that residue that becomes soil organic matter (humus) in contrast to the amount that is used as an energy source and is eventually lost out of the soil as CO_2gas.

10.2 NITRIFICATION.

Nitrification is the oxidation of NH_4^+ to NO_2^- and then to NO_3^- by chemotrophic bacteria. These bacteria use the reduced nitrogen compounds as an energy source and obtain their carbon from CO_2gas.

The two main groups of bacteria responsible for this process are nitrosomonas which carry out the oxidation of NH_4^+ to NO_2^- and nitrobacter which is responsible for the oxidation of NO_2^- to NO_3^-. The biological oxidation of ammonium to nitrite and then to nitrate is very sensitive to soil conditions such as soil pH, moisture levels, aeration and temperature.

Laboratory procedure

A. Experimental design and setup.

Each student will run one of the following experiments. The results from all three experiments will be combined and distributed to the class.

Soils code:

1 = low pH, low organic matter content.

2 = normal pH, low organic matter content.

1. Experiment 1 - effect of the C:N ratio of the added organic residue on soil nitrogen levels.

Use soil 2.

no. bottles	soil (g)	residue	mL 0.05 N KNO_3	mL H_2O
1	25	none	0	9
1	25	none	2	7
1	25	0.1 g corn cobs	0	9
1	25	0.1 g corn cobs	2	7
1	25	0.1 g alfalfa	0	9
1	25	0.1 g alfalfa	2	7

2. Experiment 2 - effect of soil pH on nitrification.

Use soil 1.

no. bottles	soil (g)	$CaCO_3$ g	0.05N $(NH_4)_2SO_4$ mL	H_2O mL
1	50	0.0	4	14
1	50	0.05	4	14
1	50	0.15	4	14
1	50	0.25	4	14
1	50	0.50	4	14

3. Experiment 3 - effect of soil moisture on nitrification.

Use soil 2.

no. bottles	soil (g)	0.05N $(NH_4)_2SO_4$ mL	ml H_2O
1	50	4	2
1	50	4	8
1	50	4	14
1	50	4	20
1	50	4	26

Store all the experiments in the laboratory drawers assigned by your laboratory instructor. All experiments (bottles) will be analyzed for nitrate nitrogen after three weeks of incubation.

B. Nitrate analysis. (Use goggles and safety clothing) At the end of the three week incubation.

a. Add one 0.5 g scoop of $CaSO_4$s and exactly 50 mL deionized water to each incubation bottle that is to be analyzed.

b. Shake intermittently for 10 minutes and let the suspension settle for an additional minute.

c. Carefully decant the liquid into a filter and collect the clear filtrate in a clean beaker.

d. Transfer 2 mL of clear filtrate into a clean 100 mL beaker.

e. Add 5 drops of saturated $Ca(OH)_2$ solution from a dropper bottle.

f. Slowly evaporate to dryness on hot plate under the hood, remove and allow to cool as soon as dry.

Note: From this point on in the laboratory safety glasses must be worn.

g. Add exactly 2 mL of phenoldisulfonic acid to the beaker. Tip the beaker so the acid contacts all of the dried precipitate.

Note! phenoldisulfonic acid is very corrosive to skin and clothing.

h. Allow the acid to react with the precipitate for 5 minutes to complex the nitrate ions.

i. Add exactly 34 mL of deionized water from a burette to the beaker and stir with a clean glass rod until all the residue is dissolved.

j. Add exactly 14 mL of 1:1 (V:V) NH_4OH to the beaker. The final solution should be yellow in color.

k. The intensity of the yellow color is related to the amount of nitrate in the soil and is determined using the UV-VIS spectrometer.

l. Your laboratory instructor will demonstrate the use of the spectrometer. Record the percentage transmission for each treatment in your assigned experiment in the following data section. Turn a duplicate of this data into your laboratory instructor.

Data - biological activity experiment.

Experiment ______

		Percentage transmission
Treatments	1	______
	2	______
	3	______
	4	______
	5	______
	6	______

Note: Experiments 2 and 3 only have five treatments.

3. Laboratory report.

Each student is responsible for a detailed report of the three biological activity experiments. Data sheets will be compiled from all of the laboratory sections and handed out in lecture by the instructor. The report should include graphs showing the amount of nitrate in the soil as a function of the experimental treatment, i.e., type of residue added (the residues C:N ratio), soil pH (amount of added calcium carbonate) and soil moisture level (amount of total water added).

The report should include a detailed discussion of the effect of the treatments on the decay and nitrification processes. The report is due your last laboratory period.

4. Answer the following questions and attach to your write-up.

a. How do heterotrophic and chemotrophic organisms differ?

b. What effect would soil aeration have on the breakdown of organic residues.

c. What soil conditions would favor nitrification?

d. What is the major form of nitrogen in soils?

e. Why was potassium nitrate used as a fertilizer in one experiment and ammonium sulfate in the other experiments?

LABORATORY 11.

SOIL TESTS FOR PHOSPHORUS AND POTASSIUM.

Soil testing is the most common method of diagnosing the fertilizer needs of plant growth. A soil test is defined as a chemical method for determining the nutrient supplying power of a soil. In theory the chemical is added to the soil as an extracting reagent. The chemical may be weak or strong acids or bases or salts of various concentrations. The extracting reagent removes a portion of the nutrient element in the soil. The level of nutrient extracted by the reagent is then determined by a standard laboratory procedure. The soil test level has little value until field research has correlated the level of the nutrient extracted by the chemical reagent with plant response. For this reason the soil test is restricted to the soils and crops that were involved in the correlation studies. If the soil test is to be extended, to new soils and/or crops, additional field studies must be performed.

11.1 Phosphorus - Bray-Kurtz P_1 TEST.

Recommendations for P fertilization in Illinois is based on: 1. The level of P extracted by the Bray-Kurtz P_1 test reagent (0.025 N HCl + 0.03 N NH_4F). 2. The native phosphorus fertility of the subsoil. Illinois has been divided into three regions in terms of the native phosphorus fertility of the soil horizons below the plow layer which are not subject to fertilization.

Figure 11-1 Phosphorus supplying power.

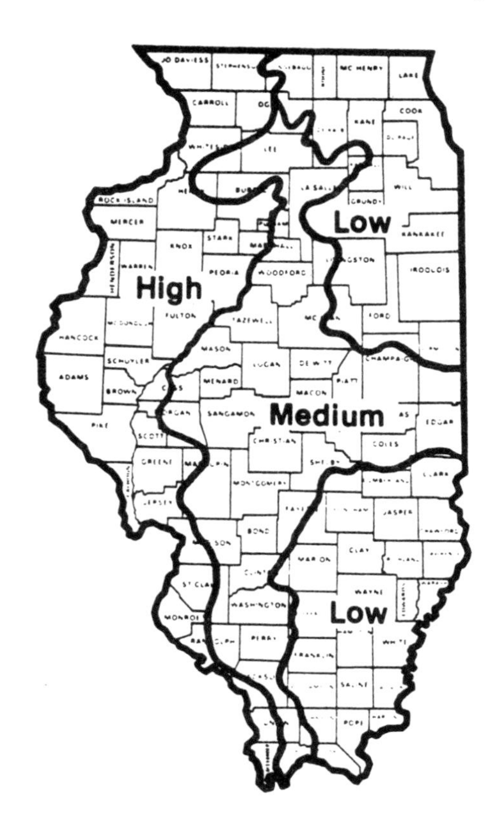

The high region corresponds to soils that formed primarily in loess that was more than 4 to 5 feet thick. The medium region corresponds to soils that formed from approximately 3 feet of loess over Wisconsin age till. While the low region corresponds to soils that formed from thin loess over either Wisconsin age tills (northern area) or Illinoian aged tills (southern area).

A high native phosphorus fertility means that the soil will have a high P_1 test and conditions that are favorable for good root growth. A low native phosphorus fertility may be caused by a number of factors including:

a. A low P_1 test due to low levels of phosphorus in the parent materials or loss of phosphorus during soil formation or fixation of the phosphorus due to high pH values.
b. Poor root growth due to poor internal drainage and the subsequent poor aeration or due to compacted zones that mechanically inhibit root growth.
c. Shallow or droughty soils.

The Bray-Kurtz P_1 test involves determining the amount of P extracted from the soil by a chemical reagent. The amount of P extracted (P_1 test value) is then used to predict the amount of fertilizer needed by a given crop to achieve the desired yield goals. The P_1 test is based on studies that have correlated the level of phosphorus extracted by the Bray-Kurtz reagent with crop response to added P fertilizer.

Interpretation - phosphorus soil test.

Figure 11-2 shows the relationship between the P_1 test of soils from the high, medium and low phosphorus supplying regions and the percent maximum yield of different crops. The figure illustrates that crops respond differently to soil phosphorus levels, yields of corn and soybeans have reached maximums at P_1 test values of 30 in the high regions, but wheat, oats, alfalfa and clover still have not reached maximum yields at P_1 test of 60.

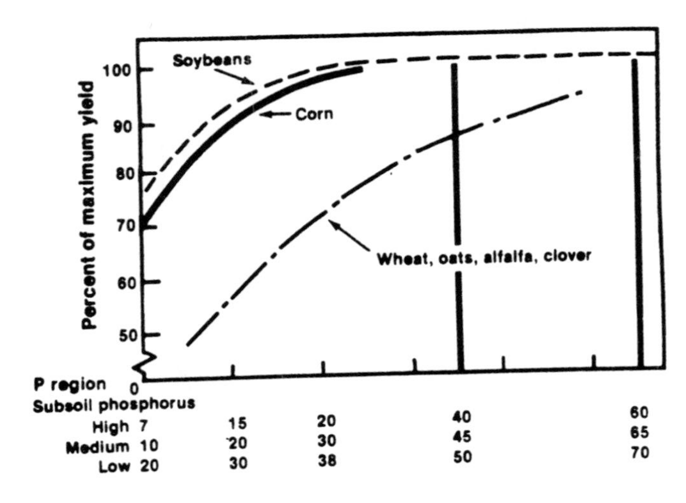

Figure 11-2 Relationship between yield and P_1 test.

Figure 11-2 also illustrates that for soils in the high, medium and low phosphorus supplying regions that the recommended minimum P_1 test values are 40, 45, and 50 and the recommended maximum P_1 test values are 60, 65, 70 respectively.

Fertilization goals in Illinois are two part; the first goal is to build the soil up to the minimum recommended P_1 test for the region over a four year period, this is called **buildup**. The second goal is to replace the phosphorus removed by the crop, this is called **maintenance.**

Buildup - P. Research has shown that, as an average 9 pounds of P_2O_5 per acre is needed to increase the P_1 test by one unit. To achieve the buildup goal nine times the difference between the minimum P_1 test for the region and the actual P_1 test of the soil must be applied over a four year period.

The soil test program in Illinois recommends testing the soil every four years to determine if buildup fertilizer needed to be applied to the soil in addition to maintenance amounts. No phosphorus either as buildup nor maintenance is recommended is soil test values exceed the recommended maximum P_1 test values. But the soil should be tested every two years to determine that soil test levels do not drop below recommended values.

The following equations can be used to calculate the P_2O_5 required to build the soils P_1 test up to the minimum value over a four year period.

$$P_2O_5 \text{ needed} = (\text{desired } P_1 \text{ test} - \text{actual } P_1 \text{ test}) \times 9/4$$

For example, if a grower has a soil in the medium phosphorus supplying region with a P_1 test of 30, the fertilizer needed each year for four years to change the P_1 test from 30 to 45 would be given by the following calculation:

$$P_2O_5 \text{ needed} = (45 - 30) \times 9/4 = 33.75 \text{ (34 lbs } P_2O_5\text{/acre)}$$

Maintenance - P. In addition to adding phosphorus fertilizer to build the soil's P_1 test value up to the minimum, additional fertilizer is added to replace the amount of phosphorus removed with the crop. Values in Table 11-1 reflect the amount of phosphorus removed by various yields of different crops. The values for wheat and oats is equal to 1.5 times the amount of phosphorus removed by the grain. This correction has been accounted for in the maintenance values for those crops.

11.2 Potassium - Ammonium acetate K test.

Illinois has been divided into two regions for potassium based on the cation exchange capacities of the soils. Soils with cation exchange capacities less than 12 $cmol_c$/kg are considered to be low, while soils with CEC's equal to or greater than 12 $cmol_c$/kg are high.

Within the regions differences in K fertility can be caused by:

a. The amount of clay and organic matter in the soil. This determines actual CEC of soil.
b. The degree of weathering of the soil. Weathering tends to leach K out of the soil profile.
c. Nature of the parent material. Wisconsin till > loess > weathered Illinoian aged till.
d. Soil conditions that effect root growth, such as drainage, compaction or adverse soil pH values.

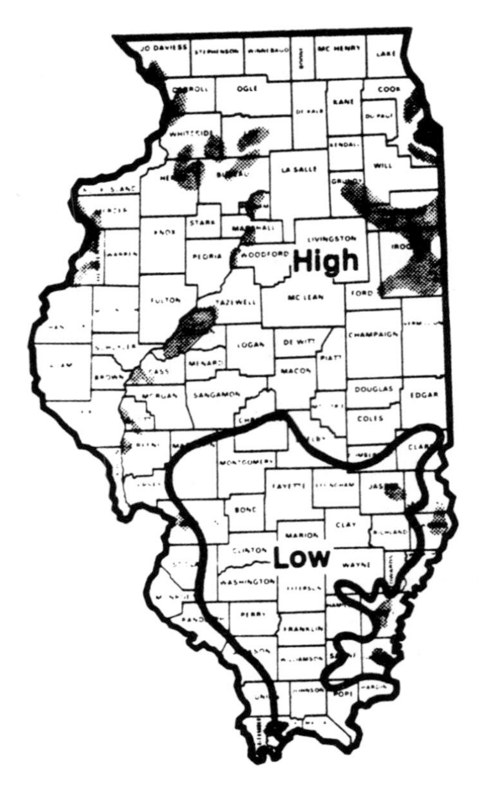

Figure 11-3 Approximate cation exchange capacities of Illinois soils.
Shaded areas are sand with low CEC.

In general the "low" areas are either sandy soils with low CEC values and hence not much ability to hold exchangeable K and soils in southern Illinois that formed from thin loess over Illinoian aged till. These soils tend to be more highly weathered than deep loess or thin loess over Wisconsin aged till soils and tend to have lower CEC values because of their clay types and organic matter contents.

Interpretation - potassium.

Figure 11-4 gives the suggested minimum and maximum K test values for soils in the high and low CEC regions. For soils in the low CEC regions soil test values should be between 260 and 360.

For soils in the high CEC region soil test values should be between 300 and 400.

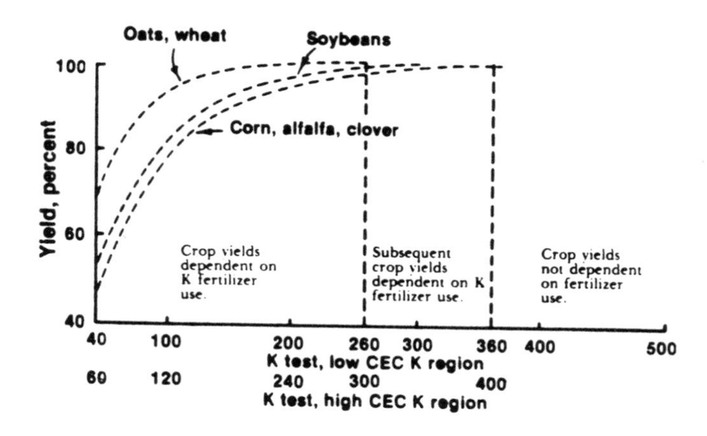

Figure 11-4 Relationship between soil K test values and yield.

On most soils potassium will remain in the soil and soil test values can be built up over a period of time as with phosphorus. On some soils soil test values for potassium do not tend to increase with build up applications of potassium. For these soils annual applications of potassium are recommended.

Producers who have one or more of the following conditions should consider the annual application option:

1. Soils for which past records indicate that soil-test potassium does not increase when build-up applications are applied.
2. Sandy soils with very low CEC values.
3. Producers who have an unknown or very short tenure arrangement for a particular farm.

On all other fields, use the build-up, maintenance option.

Build-up - K. Research has shown that 4 pounds of K_2O are required, on the average to increase the soil test 1 pound. If build-up is achieved over a four year rotation then the amount of K_2O needed is equal to the difference between the desired K test and the actual K test. Therefore, the recommendation for build-up is equal to four times the difference between the actual soil test value and the desired soil test value. Soil should be sample for potassium between May 1 and

September 30 or should be adjusted according to the procedure given in the footnote[1].

$$K_2O\text{needed} = (\text{Desired K test} - \text{Actual K test})$$

Annual application option - K. If the soil test levels for potassium are below the desired build-up goal, apply potassium fertilizer annually at an amount equivalent to 1.5 times the maintenance rate. If the soil test values are only slightly below to 100 units above desired goals then apply maintenance only.

Maintenance - K. As with phosphorus, the amount of fertilizer required to maintain the soil test values equals the amount removed by the harvested portion of the crop. Table 11-1 gives the maintenance amounts of potassium and phosphorus for various crops and yields.

---

[1] Test on soil samples that are taken before May 1 or after Sept 30 should be adjusted downward as follows: subtract 30 pounds for dark-colored soils in central and northern Illinois: 45 pounds for light-colored soils in central and northern Illinois, and fine textured bottomland soils; and 60 pounds for medium and light-colored soils in southern Illinois.

Table 11-1 Maintenance fertilizer (P and K) required for various yields of crops.

Yield, bushels or tons per acre	P_2O_5	K_2O[a]
	----- *pounds per acre* -----	
Corn grain		
90 bu	39	25
100	43	28
110	47	31
120	52	34
130	56	36
140	60	39
150	64	42
160	69	45
170	73	48
180	77	50
190	82	53
200	86	56
Oats		
50 bu	19[b]	10
60	23	12
70	27	14
80	30	16
90	34	18
100	38	20
110	42	22
120	46	24
130	49	26
140	53	28
150	57	30
Soybeans		
30 bu	26	39
40	34	52
50	42	65
60	51	78
70	60	91
80	68	104
90	76	117
100	85	130
Corn silage		
90 bu; 18T	48	126
100; 20	53	140
110; 22	58	154
120; 24	64	168
130; 26	69	182
140; 28	74	196
150; 30	80	210
Wheat		
30 bu	27[b]	9
40	36	12
50	45	15
60	54	18
70	63	21
80	72	24
90	81	27
100	90	30
110	99	33
Alfalfa, grass, or alfalfa-grass mixtures		
2T	24	100
3	36	150
4	48	200
5	60	250
6	72	300
7	84	350
8	96	400
9	108	450
10	120	500

[a] If the annual application option is chosen, then K application will be 1.5 times the values shown below.
[b] Values given are 1.5 times actual removal.

Laboratory procedures.

1. Determine the P_1 and K test values for the soils provided by the instructor. Based on your results, the area of the Illinois that the soils are from as well as the crop(s) to be grown and expected yield(s) make a fertilizer for:

   a. Continuous corn.
   b. Continuous soybeans.
   c. Corn-soybean rotation.
   d. Corn-soybean-wheat-alfalfa-alfalfa rotation.

Your lab instructor will provide information on soil type, location and expected yield goals for the crops.

Procedure Bray-Kurtz P_1 test.

1. Record the exact weight of a 1 gram level scoop of soil.

2. Place the soil in a clean dry 20 mL test tube.

3. Add 10 mL of the PA (phosphorus extracting solution; 0.025 N HCl + 0.03 N NH_4F) to the test tube, cover with a clean stopper or parafilm and shake for **exactly** one minute.

4. Filter immediately through Whatman No. 2 filter paper.

5. Pipet 5 mL of **clear** filtrate into a clean test tube.

6. Add 5 drops of PB (ammonium molybdate + HCl in boric acid) solution to the 5 mL of clear filtrate and mix by swirling for 1 minute.

7. Add 5 drops of PC (amino-naphthol-sulfonic acid) solution to the test tube and swirl to mix.

8. Fifteen minutes after adding the PC solution, determine the concentration of P (lbs/acre) in the sample using the spectrometer and standard curve provided by your instructor.

Data for P_1 test

	Percentage transmission	lbs P_2O_5 needed per acre
Soil 1	______	______
Soil 2	______	______

Procedure ammonium acetate K test.

1. Record the exact weight of a 1 gram scoop of soil.

2. Place soil in a clean 20 mL test tube.

3. Add 10 mL of 1 N ammonium acetate, cover with a clean stopper or parafilm and shake for 5 minutes.

4. Filter through Whatman no. 2 filter paper into a clean vial.

5. Determine K in the extract by emission spectroscopy.

Note. Instructor will furnish standard curve for both soil tests (P_1 test and K test).

Data for K test

	Absorbance	lbs K_2O needed per acre
Soil 1	______	______
Soil 2	______	______

2. Laboratory report should be written in the form of a letter to a farmer detailing the results of the soil tests for P and K and the recommendation for the rotations.

3. Answer the following questions:

   a. What is a soil test?

   b. Why isn't total soil K or P used to determine the fertility status of the soil?

   c. What soil properties determine the native K and P fertility of Illinois soils?

   d. Why isn't the Bray-Kurtz P1 test used in the western U.S. on calcareous soils?

   e. What soil conditions result in the fixation of added K fertilizer?

   f. Discuss the fate of added fertilizer P in acid, neutral and calcareous soils.

   g. Are P_2O_5 and K_2O the forms of fertilizer applied to the soil? If not what are the common forms of these nutrients that are applied to soils?

   h. Why is there no suitable soil test for determining nitrogen fertilizer rates?

i. What is "buildup" and "maintenance" refer too in the Illinois fertility program?

j. Why do soils with CEC values greater than 12 $cmol_c/kg$ require more K fertilizer than soils with lower CEC values.